Strategic Priorities for Agricultural Development in Eastern and Central Africa

Steven Were Omamo
Xinshen Diao
Stanley Wood
Jordan Chamberlin
Liangzhi You
Sam Benin
Ulrike Wood-Sichra
and
Alex Tatwangire

RESEARCH REPORT 150

INTERNATIONAL FOOD POLICY RESEARCH INSTITUTE
sustainable solutions for ending hunger and poverty
IFPRI®

ASARECA
THE ASSOCIATION FOR STRENGTHENING AGRICULTURAL RESEARCH IN EASTERN AND CENTRAL AFRICA

International Food Policy Research Institute
2033 K Street, NW
Washington, D.C. 20006-1002, U.S.A.
Telephone +1-202-862-5600
www.ifpri.org

DOI: 10.2499/9780896291584RR150

Library of Congress Cataloging-in-Publication Data

Strategic priorities for agricultural development in Eastern and Central Africa / Steven Were Omamo . . . [et al.].
p. cm. — (Research report ; 150)
Includes bibliographical references.
ISBN-13: 978-0-89629-158-4 (alk. paper)
ISBN-10: 0-89629-158-8 (alk. paper)
1. Agriculture—Economic aspects—Africa, Eastern. 2. Agriculture—Economic aspects—Africa, Central. 3. Agricultural development projects—Africa, Eastern. 4. Agricultural development projects—Africa, Central. I. Omamo, Steven Were. II. International Food Policy Research Institute. III. Series: Research report (International Food Policy Research Institute) ; 150.
HD2118.S76 2006
338.10967—dc22 2006032181

Contents

Tables

Figures

Box

Foreword

In countries that are heavily dependent on agriculture for employment and income, underperformance is not only untenable but also potentially explosive. This is the case in the countries of eastern and central Africa—Burundi, Democratic Republic of Congo, Eritrea, Ethiopia, Kenya, Madagascar, Rwanda, Sudan, Tanzania, and Uganda—where tens of millions of people face ongoing poverty, hunger, and malnutrition. This report, the result of a two-year collaboration between the International Food Policy Research Institute and the Association for Strengthening Agricultural Research in Eastern and Central Africa, identifies how eastern and central African countries can stimulate agricultural growth to address these dire circumstances.

The findings suggest that improved agricultural performance will require investments that foster productivity growth, strengthen markets, improve rural linkages between the agricultural and nonagricultural sectors, and promote regional cooperation. Of particular interest is the identification of the most performance-enhancing commodity subsectors, in an economy-wide setting, and the "agricultural development domain" singled out as most promising for targeted investment.

These results and their implications are being widely discussed and debated in the countries of eastern and central Africa, in many cases shaping policy and investment strategies. We hope that the findings, made available through this report, will elicit similar responses in other regions.

Joachim von Braun
Director General, IFPRI

Seyfu Ketema
Executive Director, ASARECA

Acknowledgments

This report is one of the key outputs of a collaborative project between the Association for Strengthening Agricultural Research in Eastern and Central Africa (ASARECA) and the International Food Policy Research Institute (IFPRI) titled "Strategies and Priorities for Sub-Regional Agricultural Development and Sub-Regional Agricultural Research-for-Development in Eastern and Central Africa." A steering committee drawn from the ASARECA Secretariat, ASARECA's Eastern and Central Africa Programme for Agricultural Policy Analysis (ECAPAPA), and IFPRI guided project implementation. Several of technical advisory groups were created to cover specific issues ranging from data collection and integration to regional agricultural sector modeling. These technical advisory groups helped promote information sharing, consultation, buy-in, and collective ownership of the process among key stakeholders. Members of the technical advisory groups were drawn from national agricultural research institutes, universities, CGIAR centers, nongovernmental organizations, and the private sector. Funding for the project came from the United States Agency for International Development's Regional Economic Development Support Organization (USAID-REDSO). All of these important contributions are gratefully acknowledged.

Summary

Agricultural development strategies delineate priorities for actions to enhance agricultural and overall development. They are usually put forward by individual countries based on assessments of national needs. Seldom are attempts made to identify strategic priorities for agricultural development that cut across national boundaries. This gap is perhaps not surprising—organizations mandated to develop and implement regional agricultural development programs are rare. Although the gap may be understandable, it is also troubling.

This report helps to fill that gap for eastern and central Africa (ECA), focusing on Burundi, Democratic Republic of Congo (DRC), Eritrea, Ethiopia, Kenya, Madagascar, Rwanda, Sudan, Tanzania, and Uganda. Recent trends and the current performance of agriculture in these countries expose a region progressively less able to meet the needs of its burgeoning population. With agriculture looming so large in most ECA economies, sluggish growth in agricultural productivity has translated into sluggish overall growth and generally low per capita income levels. High levels of agricultural importation—particularly of staples—appear to be only partially filling the consumption needs of a population lacking purchasing power, resulting in extensive adult and child malnutrition and towering child mortality rates.

Such forces as globalization, market liberalization, privatization, urbanization, HIV/AIDS, population growth, climate change, and the changing proprietary nature of agricultural technology are redefining many of the problems facing agricultural policymakers in ECA, and thus the kinds of policy solutions required. Most of these forces have roots and expressions that extend beyond national boundaries, implying the need for broad perspectives and regional responses. Neighboring countries might gain from cooperating in key areas of agricultural development. This report is motivated by such regional potentials in ECA.

The analytical approach is explicitly strategic. First, using geographic information systems methods to identify and depict spatial similarities and differences in the context of agriculture in ECA, the analysis spans all 10 countries in the region, thereby permitting simultaneous focus on both national and regional phenomena. Agricultural development domains representing particular realizations of agricultural potential, access to markets, and population density are used to help highlight differences and similarities in agricultural development priorities and options across the region. Second, using a dynamic economic model of agriculture in ECA, known as a multimarket model, the analysis includes numerous agricultural and nonagricultural subsectors while tracking broader economic conditions in a forward-looking setting. Third, using a model that quantifies the effects of productivity-enhancing investments in agricultural research and development (R&D), known as the Dynamic Research Evaluation for Management (DREAM) model, the analysis explores the potential returns to regional cooperation in agricultural development.

To build understanding of the strategic opportunities for agricultural development in ECA, the implications for overall economic growth and poverty reduction of alternative scenarios of agricultural growth are examined using the multimarket model. A central piece of the analysis is a business-as-usual scenario that projects recent trends in agricultural growth into the future. The business-as-usual scenario therefore serves as a base against which to evaluate alternative agricultural development strategies for ECA.

Business-as-usual outcomes suggest that in all countries except Sudan and Uganda (assumed to continue to register relatively high growth rates as they recover from civil strife), agricultural gross domestic product (AgGDP) and overall gross domestic product (GDP) would grow at rates below the 3 percent required to keep pace with population growth. Per capita GDP growth rates would therefore stand at below 1 percent in a majority of countries. Kenya's per capita GDP growth to 2015 would be essentially zero; those of Madagascar, Rwanda, and Tanzania would be only marginally higher. Burundi, DRC, Eritrea, and Ethiopia would register negative per capita GDP growth rates.

Clearly, with business-as-usual in agriculture, ECA's future would not feature broad-based economic growth. Not a single ECA country would achieve the estimated 6 percent GDP growth rate required to meet the United Nations Millennium Development Goal (MDG) of halving poverty by 2015. Other development goals identified by ECA countries—such as increased food and nutrition security—would also remain beyond reach. The gap between demand and supply of major food crops in ECA would widen. For cereals, the supply shortfall would increase to 6 million metric tons by 2015, 50 percent more than that in 2003, and 15 percent of total regional demand.

Further analysis with the multimarket model yields numerous insights into the nature of agricultural development that might allow countries to avoid business-as-usual outcomes:

- Achieving GDP growth rates required to meet MDG poverty reduction targets would imply threefold increases in agricultural sectoral and subsectoral growth rates.
- Whereas growth in export subsectors is often put forward as a pathway out of poverty for countries in ECA, the analysis reveals that the largest poverty reductions would come from growth in subsectors for which demand is greatest within the region—such as staples, livestock products, oilseeds, and fruits and vegetables. Increasing productivity in these subsectors would directly benefit the great majority of ECA's numerous small farmers by easing key resource constraints in the activities to which they devote most of their resources.
- When ECA is viewed as a region, milk emerges as the most important commodity subsector for growth-inducing investment in agriculture, based on simulated cumulative contributions to overall GDP to 2015. Oilseeds, cassava, and fruits and vegetables also rank highly. Viewed together, staples subsectors result in the largest GDP gains, followed by livestock products, fruits and vegetables, and oilseeds.
- Priorities for Kenya, Tanzania, and Uganda closely match those of the region. Regional priorities appear to be less relevant for Eritrea, Ethiopia, and Sudan, with cereals and milk being more important in these countries than in the others. Regional priorities appear to have little relevance for Burundi, DRC, Madagascar, and Rwanda. In Burundi and Rwanda, bananas, potatoes, and sweet potatoes are crucial. In DRC, oilseeds and cassava are paramount. In Madagascar, rice is central.
- Whereas growth in the cassava and milk subsectors generate the largest aggregate gains, such gains would be concentrated in a handful of countries. Fruits and vegetables, beef, oilseeds, and maize emerge as commodity subsectors in which growth would yield gains that were both large and widespread.
- Balanced growth strategies featuring growth in several agricultural subsectors lead to higher overall economic growth than does that featuring growth in a few sectors.
- Agricultural productivity growth alone is insufficient to meet MDG poverty reduction targets. Growth in nonagricultural sectors and improvements in market conditions are also required.

- Because poverty rates vary geographically within countries, growth strategies that take such differences into account lead to larger reductions in poverty than do those that ignore such variations.
- The agricultural development domain characterized by high agricultural potential, low market access, and low population density (HLL; see Table 3.8 for the definitions of the various domains) emerges as the clear priority for efficient, equitable, and sustainable growth in the region. The greatest scope for broad-based benefits from regionally conceived initiatives in agricultural development resides primarily in this domain. That scope appears to be substantial. Agriculture-based growth in the LLL, HHH, and HLH domains is also important and probably offers scope for both poverty reduction and benefits from regional cooperation. But such potential is likely to be more difficult to achieve. Agriculture-based growth in the LHH, HHL, LLH, and LHL domains is unlikely to be large enough to warrant major investments in agricultural development. Best-bet growth-enhancing options in these areas probably lie outside agriculture.
- Using agricultural R&D as an illustration, significant returns to regional cooperation in agricultural development are identified.

The analysis therefore suggests that to avoid the bleak growth and poverty outcomes implied by business-as-usual in agriculture, ECA governments must invest in combinations of measures that (1) spur productivity growth, focusing on subsectors with high demand within ECA; (2) strengthen agricultural markets; (3) enhance linkages between agricultural and nonagricultural sectors; and (4) exploit opportunities for regional cooperation.

CHAPTER 1

Introduction

Agricultural development strategies outline priorities for actions to achieve enhanced agricultural and overall development. They are usually put forward by individual countries, based on assessments of national needs. Seldom are attempts made to identify strategic priorities for agricultural development that cut across national boundaries. This gap is perhaps not surprising. Organizations mandated to develop and implement regional agricultural development programs are rare. The gap may be understandable, but it is also troubling.

Such forces as globalization, market liberalization, privatization, urbanization, HIV/AIDS, population growth, climate change, and the changing proprietary nature of agricultural technology are redefining many of the problems facing agricultural policymakers and thus the kinds of policy solutions required. Most of these forces have roots and expressions that extend beyond national boundaries, implying the need for broad perspectives and regional responses. Neighboring countries might gain from cooperating on key negotiating problems with the rest of the world. For example, a regional bloc might achieve greater negotiating power and leverage than would several countries acting individually in dealing with the World Trade Organization or with other regional groupings. Some countries in a region might be able to act as regional growth centers and pull neighboring countries along with them as they grow. For example, they might buy imports from their neighbors, attract migrant workers, and be sources of investment capital. These regional trade dynamics can be more powerful if key development policies are synchronized across countries. Finally, some national investments might generate benefits for their neighbors, leading to efficiency gains from regional rather than national investment strategies. For example, agricultural research and development (R&D) in one country might lead to spillover benefits for neighboring countries that have similar agroecological conditions. It might be inefficient for each country to undertake wholly independent R&D; significant gains might be achieved from regionally conceived and implemented R&D programs.

This report is motivated by such regional potentials in 10 countries in eastern and central Africa (ECA): Burundi, Democratic Republic of Congo (DRC), Eritrea, Ethiopia, Kenya, Madagascar, Rwanda, Sudan, Tanzania, and Uganda (Figure 1.1).[1] Such a motivation would

[1]This geographic coverage is defined by the mandate of the principal partner in the project within which the analysis was undertaken, namely, the Association for Strengthening Agricultural Research in Eastern and Central Africa (ASARECA). ASARECA's regional mandate covers these 10 countries. Clearly, the approach to grouping countries and subnational regions within them is crucial to the type of analysis presented here. Any choice will be arbitrary to some extent. The existence of a body with a regional mandate, such as ASARECA, is critical to the decision on which countries to include in the analysis. As is shown in Chapter 3, using spatial modeling methods to define analytical units and demarcate potential intervention locales within the region is highly productive.

Figure 1.1 The study region: Eastern and central Africa

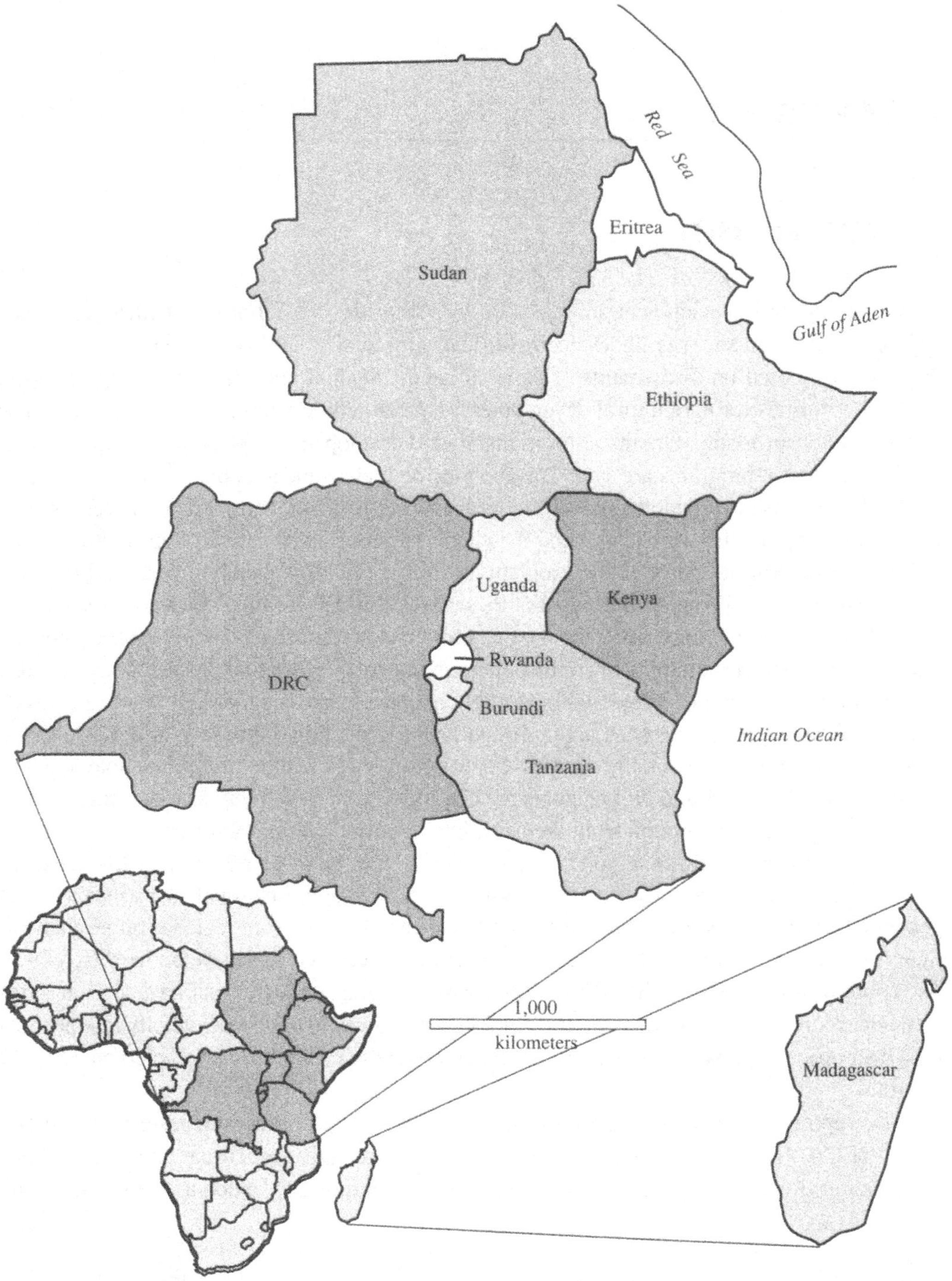

appear to fit with recent political and economic developments in the region and elsewhere in Africa. The East African Community has been resurrected, aiming to widen and deepen cooperation between the three member states of Kenya, Tanzania, and Uganda in political, economic, and social fields for mutual benefit (EAC 2004). The Inter-Governmental Authority for Development, covering Djibouti, Eritrea, Ethiopia, Kenya, Somalia, Sudan, and Uganda, is becoming increasingly important in key areas, most notably conflict resolution (IGAD 2004). The New Partnership for Africa's Development (NEPAD) is convening major regional initiatives, including the Comprehensive Africa Agriculture Development Program, which aims to sensitize African policymakers about the need for concerted action toward sustainable agricultural development in Africa (NEPAD 2004). The 1991 Abuja Treaty establishing the African Economic Community designated the regional economic communities (RECs) as the lead agents for realizing the dream of a pan-African economic community. In 2003, The Heads of State of the NEPAD Implementation Committee have assigned the RECs priority tasks to be carried out to advance the NEPAD agenda. Also in 2003 at the African Union Summit in Maputo, African heads of state and government signed the Maputo Declaration on Agriculture and Food Security, in which these leaders committed their governments to allocating at least 10 percent of national budgetary resources to agricultural sectors (SADC 2004). These developments render this an especially opportune moment to consider agricultural development in a regional context.

A central premise of this report is that different agricultural development priorities imply different patterns of investment within agricultural sectors, which, in turn, result in different effects on growth within agricultural sectors, growth in wider economies, and overall poverty rates. The analysis in the report therefore tracks and quantifies such linkages. A second important premise is that the generally poor recent performances of agricultural sectors and overall economies in ECA signal poorly conceived agricultural development strategies. The analysis therefore aims to provide a basis for improved agricultural development strategy formulation in the region.

A third premise, closely related to the second, is that an agricultural development strategy exists even when it is not made explicit—in the form of extant policies and institutions that define opportunities and constraints in agricultural sectors. The more coherent and transparent the strategy, the clearer will be the signals it sends to sector participants as they develop their plans and set their priorities, and thus the more effective and efficient will be the associated allocations of resources. An underlying argument in the report is therefore that past agricultural development efforts may have failed to achieve their aims partly because they were based on insufficiently coherent and transparent agricultural development strategies. The analysis therefore aims to demonstrate the nature of coherent and transparent explicit agricultural development strategies.

National Agricultural Development Objectives in ECA

The 10 countries under consideration cover an area of 8.5 million km^2 with a total population of more than 280 million people, most of whom are rural dwellers pursuing agricultural livelihoods (Tables 1.1 and 1.2). The 10 countries have different social, political, and economic histories, and thus also distinct legal and institutional structures and processes. Despite such differences, however, there is considerable similarity across countries in factors viewed to constrain agricultural development, and thus also in agricultural policy objectives.[2]

Agricultural policy in Burundi focuses on enhancing productivity and reducing pressure on land through improved access to key factors of production, diversification of agricultural exports, and better integration of crop and livestock production.

When peace returns to DRC, the central agricultural policy challenges will revolve around rebuilding decimated agricultural

[2]A full description of national agricultural policy regimes in ECA, including policy objectives and plans for the future, can be found in Ngigi (2004).

Table 1.1 Total population in ECA countries, 1962–2002 (000s population)

Country/region	2000	2002	2005	2010	2015	Percentage of ECA
Burundi	6,267	6,602	7,319	8,631	9,834	2.45
DRC	48,571	51,201	56,079	64,714	74,160	18.71
Eritrea	3,712	3,991	4,456	5,256	5,914	1.48
Ethiopia	65,590	68,961	74,189	83,530	93,845	24.51
Kenya	30,549	31,540	32,849	34,964	36,864	10.58
Madagascar	15,970	16,916	18,409	21,093	24,000	6.12
Rwanda	7,724	8,272	8,607	9,559	10,565	2.84
Sudan	31,437	32,878	35,040	38,323	41,430	11.37
Tanzania	34,837	36,276	38,365	41,931	45,909	12.52
Uganda	23,487	25,004	27,623	32,996	39,335	9.42
ECA	268,144	281,641	302,936	340,997	381,856	100.00
Developing countries	4,754,076	4,899,943	5,117,471	5,478,810	5,832,660	6.04
Sub-Saharan Africa	609,779	685,071	687,513	770,166	857,638	43.64

Source: FAO n.d. (accessed 2004).

institutions, ranging from commodity markets to research and extension systems. The aim will be to transform the economy away from its historically heavy dependence on raw commodity exports toward increased domestic addition of value.

Compared with that of other countries in ECA, agriculture's share of gross domestic product (GDP) in Eritrea overall is low (24 percent versus an average of 43 percent for the region).[3] Yet 80 percent of the county's population pursues agriculture-based livelihood strategies. The government's broad goals for the sector are increased food supplies and enhanced productivity and commercialization in smallholder agriculture, especially through increased irrigation.

Ethiopia's overall economic development is considered to hinge on agricultural development in agricultural development–led industrialization. This type of development envisions intensification of agriculture, deepening of technology, and commercialization of smallholder agriculture as essential components of enhancing the sector's capacity to drive the economy.

Kenya's newly formulated Strategy to Revitalize Agriculture signals a new proactive public policy stance toward agriculture. In an apparently deliberate turn away from the largely disappointing market-driven approach of the 1990s, toward the mixed-economy approach of the 1960s, 1970s, and early 1980s, key public institutions are being revitalized and accorded high-profile roles in spurring growth, increasing food security, and reducing poverty.

Food insecurity is identified as a major impediment to agricultural and overall economic development in Madagascar. Improved natural resource management and modernization of farming practices are viewed as crucial to increasing food security and fully exploiting Madagascar's favorable natural endowments.

Agricultural policy in Rwanda focuses on agriculture's contribution to the overall policy goals of increasing and diversifying exports, reducing structural trade deficits, and building foreign exchange reserves. Within the sector, replenishing soil fertility and increasing use of high-yielding technologies are priorities.

[3]Some estimates put this share as low as 12 percent. Services and, to a lesser extent, industry account for the bulk of the country's GDP.

Table 1.2 Share of rural population in ECA countries, 1962–2002 (%)

Region/country	2000	2002	2005	2010	2015
Burundi	91.03	90.40	89.40	87.53	85.49
DRC	69.72	68.86	67.32	64.04	60.73
Eritrea	81.28	80.48	79.24	76.77	73.82
Ethiopia	84.48	83.79	82.66	80.51	78.01
Kenya	66.63	64.65	61.74	57.08	52.83
Madagascar	70.51	69.23	67.27	63.93	60.56
Rwanda	93.85	93.60	93.17	92.27	91.09
Sudan	63.88	62.03	59.28	55.05	51.32
Tanzania	67.75	65.62	62.56	57.85	53.82
Uganda	85.84	85.12	83.96	81.79	79.31
ECA	74.85	73.69	71.90	68.88	65.95
Developing countries	59.91	58.81	57.10	54.31	51.57
Sub-Saharan Africa	67.61	64.73	64.47	61.39	58.44

Source: FAO n.d. (accessed 2004).

Agricultural policy formulation in Sudan has been hampered by protracted civil strife. Major constraints to agricultural development and targeted for policy action include the build-up of weeds, pests, and diseases in irrigated areas; soil degradation; and deterioration of water resources.

Agriculture's contribution to poverty reduction and overall growth in Tanzania is highlighted by an ambitious agricultural sector growth target of 6 percent per year, accompanied by plans to create an enabling environment for agricultural productivity growth featuring new public-private partnerships.

Uganda's comprehensive and multisectoral Plan for Modernization of Agriculture aims to turn agriculture into a profitable, competitive, sustainable, and dynamic primary and agro-industrial enterprise.

Five agricultural development objectives emerge as regionally crosscutting: alleviating poverty, promoting food and nutrition security, promoting commercialization of smallholder agriculture, generating foreign exchange, and increasing agricultural production and productivity.[4] Such well-reasoned and clearly articulated agricultural development objectives are not new in ECA. Countries have been including such objectives in national and agricultural development plans for decades. Unfortunately, as described in later chapters, these objectives have remained largely unmet. Poverty remains high, food and nutrition security low, commercialization and foreign exchange generation in smallholder agriculture limited, and, most crucially, overall agricultural productivity low. What are the implications of the continuation of such conditions in ECA agriculture? Which investment strategies might lead the region away from a future driven by such outcomes? How might ECA countries attain aggregate and sectoral growth rates required to achieve such goals as the United Nations Millennium Development Goal to halve poverty rates by 2015? Providing answers to these and a range of related questions drives to the core of this document.

Study Approach

The analytical perspective is explicitly forward-looking and strategic, providing

[4]One might argue that increasing agricultural production and productivity could be subsumed in some of the other objectives. But country after country identifies it as an agricultural policy objective in its own right.

regional and national agricultural development policymakers and stakeholders with a frame of reference for planning for and investment in sustainable agricultural development in ECA. The strategic perspective implies that athough the many problems and challenges facing agriculture in ECA are identified and analyzed, much greater attention is given to the search for solutions.

The study has five distinguishing features that set it apart from most development strategies currently in place in the region (Ngigi 2004). First, available evidence and data are compiled and analyzed to address the issues. Where appropriate, results from recent studies on related topics are incorporated. Second, where feasible and profitable, empirical research is undertaken to generate new insights. Third, findings are documented as rigorously as possible, but at a level of clarity appropriate for a primarily nontechnical audience. Fourth, although such a report cannot fully address the range of important issues pertaining to policy implementation, findings are translated into a set of recommendations for consideration by relevant national and regional bodies, and issues of operational feasibility are addressed in light of political and institutional conditions. Finally, to improve its accessibility and raise prospects for its use by busy policymakers and other leaders, this report is relatively brief.

The core of the analysis focuses on benefits of various investment and policy changes in agriculture by quantifying linkages among improved conditions and incentives for production, incomes, and welfare. However, the analysis does not account for costs—such as those associated with developing new technologies, building roads, and implementing new policy. Incorporating such costs would lead to more nuanced policy conclusions, as discussed below. Lack of reliable data and daunting analytical challenges rendered such a treatment impossible. However, as is argued in the concluding chapter, anecdotal evidence suggests that certain policy recommendations may be more plausible than others based on cost considerations.

The report does not explicitly address the effects of two phenomena of extreme importance in ECA, namely, HIV/AIDS and civil strife. Unless ECA countries come fully to grips with HIV/AIDS and its implications for all development policy—including that in agriculture—little strategic value will be derived from exercises such as the current one. Similarly, unless sustainable solutions can be found to the civil strife affecting significant segments of the region's population, even the best and most competently implemented development strategies will be continually undermined by the social, political, and economic instability that invariably accompanies war. Even in the absence of permanent solutions to these two burdens, countries face the immediate task of putting in place policies that seek to promote agricultural development for economic growth and poverty reduction. Countries with coherent agricultural development strategies may be better able to recognize and integrate sustainable responses to HIV/AIDS than can those without; they may also be better prepared to seize opportunities opened up by hard-won peace.

Outline of the Report

As noted above, the question at issue in this report is the nature of agricultural development strategies that might lead ECA countries toward development paths that feature sustainable increases in agricultural productivity, food and nutrition security, and poverty reduction. To address these issues, the remainder of the report is organized as follows. Chapter 2 outlines the analytical approach, detailing the spatial and economic modeling exercises that underpin the analysis. Chapter 3 describes the socioeconomic and biophysical underpinnings of agriculture in the region, focusing on the implications of population density, agroecological conditions and agricultural potential, and access to markets, both respectively and in tandem. Agricultural development domains, representing particular realizations of population density, agricultural potential, and access to markets, are identified and various agricultural development options associated with the domains are proposed. Chapter 4 describes recent trends and current conditions in ECA agriculture. Key features of agriculture in the region are described, focusing on the contribution of agriculture to national and regional GDPs; agricultural productivity; and agricultural production, consumption, and trade. Chapter 5 explores the implications for future economic growth and poverty reduction in ECA of alternative policy and investment strategies in agriculture. Using a specially developed regional multimarket model of agriculture in ECA, outcomes from the continuation of recent trends (that is, business-as-usual) are contrasted with those associated with alternative growth-enhancing, poverty-reducing investment strategies. Chapter 6 proposes priorities for agricultural development in two dimensions: priorities among commodities for ECA as a region and for countries and groups of countries; and priorities across agricultural development domains. The potential benefits from regionally coordinated agricultural R&D initiatives are quantified. Chapter 7 contains the study's recommendations and policy implications. Broad conclusions round out the report in Chapter 8.

CHAPTER 2

Analytical Approach

The report's strategic perspective is built on three features of the analytical approach. First, using geographic information systems (GIS) methods, the analysis spans all 10 countries in eastern and central Africa (ECA), thereby permitting focus on both national and regional phenomena. Second, using a dynamic economic model of agriculture in ECA known as a multimarket model, the analysis accounts for numerous agricultural and nonagricultural subsectors while tracking broader economic conditions in a forward-looking setting. Third, using a model that quantifies the effects of productivity-enhancing investments in agricultural research and development (R&D), known as the Dynamic Research Evaluation for Management (DREAM) model, the analysis explores the potential returns to regional cooperation in agricultural development. This chapter describes the GIS methods, the multimarket model, and the DREAM model. The aim is not to detail all technical matters pertaining to these analytical tools, but rather to build understanding of why these approaches were taken and what was gained from their application. Details about all three frameworks are provided in the appendixes to this report.

Spatial Analysis Using GIS Methods

Formulating and evaluating agricultural development strategies for a region as large and diverse as ECA is extremely challenging, requiring multiple perspectives and judicious simplification. One approach involves gaining a better appreciation of the regional patterns of agriculture and of agricultural development challenges and opportunities using GIS tools and databases. Visualizing similarities and differences in agriculture across the region is a powerful means of focusing attention on areas and issues that cross national borders.

Many types of spatial analysis and mapping are feasible. The current analysis focuses on just two perspectives. First, the spatial extent, distribution, and intensity of cropland and rangeland across the region are illustrated, juxtaposed with some key regional resource and infrastructure features. Second, the region is disaggregated into geographical units (termed development domains) in which similar agricultural development problems or opportunities are likely to occur.[5]

A key goal is to use a single set of domain criteria and to apply them consistently across the region. Only with such a consistent approach can the true similarity or dissimilarity of conditions existing in, say, the highlands of Tigray in Ethiopia, be properly compared and

[5] From the national and regional strategic and policy perspectives, development domains represent areas of broadly similar strategic and investment opportunities. From a farm or enterprise perspective, development domains offer a way of identifying viable sets of livelihood options.

contrasted with those in, for example, Kenya, Tanzania, and Madagascar. These development domains permit consideration of the following issues: Where are those geographic areas within and across countries in ECA in which development problems and opportunities are likely to be most similar? Where will specific types of development policies, investments, livelihood options, and technologies likely be most effective? For established developmental successes in any given location in (or beyond) ECA, where can similar conditions be found in the region?[6]

The analysis is therefore most concerned with the geographies of attributes that constrain or enable different options for agricultural development. Based on empirical research findings both within and beyond ECA, the three specific attributes used for defining development domains are agricultural potential, market access, and population density. Although the agricultural potential of any location is a strong indicator of its absolute advantage in agricultural production, the extent to which this might actually be realized—that is, its comparative advantage—is conditioned by other factors, of which market access and population density have been shown to be significant (Pender, Place, and Ehui 1999).

Beyond mapping development domains assembled on the basis of these three factors, empirical evidence is used to explore which specific strategies are both feasible and advantageous in each domain. The locations and types of opportunities identified by this spatial analysis approach are compared with results of the economic analysis described in the following section. The economic analysis yields insights into agricultural and overall economic implications of alternative agricultural investment strategies at regional, national, and subnational levels. The complementary role of the domain analysis is to provide a visual, regional basis for examining where such investments may be most appropriately targeted. Some domains may physically span country boundaries, whereas others may manifest themselves as distinct areas within individual countries. Each domain category is defined consistently across the region. As will become clear, this consistency allows identification of truly regional agricultural development strategies.

Data used in the spatial analysis are drawn from a wide variety of secondary sources. Satellite-based interpretations of topography and land cover are from the Global Land Cover 2000 Project, the U.S. National Geospatial-Intelligence Agency, and the U.S. National Aeronautics and Space Administration. Population density and human settlement data come from the Center for International Earth Science Information Network and the International Food Policy Research Institute (IFPRI). Road infrastructure data are from the U.S. National Imagery and Mapping Agency and IFPRI. Spatially interpolated rainfall and climate station data were obtained from the University of East Anglia. Regional soil and protected area maps were compiled and harmonized from national sources by the United Nations Food and Agriculture Organization (FAO) and the United Nations Environment Programme World Conservation Monitoring Centre. Biophysical crop suitability information is from the International Institute for Applied Systems and FAO.

Economywide Multimarket Modeling

The fundamental aim of economywide multimarket models is quantification of the economic implications of alternative policy decisions or scenarios. They do so by

[6]Because national boundaries often demarcate key institutional changes, this analytical framework potentially opens scope for cross-border institutional comparisons—for example, northern Mozambique, southern Malawi, and eastern Zambia or western Kenya and eastern Uganda. Such comparisons fall outside the scope of the current analysis but would be useful areas for further research.

quantifying direct effects on supply, demand, and trade of commodities in several interlinked markets, and, where possible, by estimating the effects on household incomes of these market changes.

Most multimarket models focus on particular segments of economies. The model developed for this study focuses on agriculture but puts the agricultural sector in an economywide context. The model includes the following 33 agricultural commodities and 15 commodity groups: cereals (maize, rice, wheat, sorghum, barley, millet, oats, other cereals), root crops (potatoes, sweet potatoes, cassava, other root crops), pulses (beans, peas), oil crops (groundnuts, sesame seed, other oil crops), vegetable oil, sugar, vegetables, bananas, fruits, beverages (coffee, tea, other beverages), fiber crops (cotton), meat (bovine meat, goat and mutton meat, poultry, other meat), eggs, milk, and fish. The model also includes two aggregated nonagricultural sectors, thereby permitting capture of linkages to other segments of national and regional economies.[7]

National agricultural production, consumption, and trade data are from FAO; nonagricultural data are from the World Bank's Word Development Indicator series. Employing GIS information and methods, a range of economic data is further disaggregated. The model therefore permits analysis at multiple levels: regional, national, and subnational.

Integration of biophysical and socioeconomic information occurs at the subnational level. For each ECA country, 12 potential subnational areas are defined, based on combinations of agricultural potential (high or low), the presence or absence of irrigation (yes or no), and farmland size (small, medium, and large). Further details on the classification scheme are provided in Chapter 3.

The production side of the model is based on subnational information on the spatial distribution of agricultural production for all 32 commodities mentioned above. National production is derived by summing up the subnational production numbers.

The consumption side of the model is based on national information on commodity demand for key commodities broken down by population and income levels and disaggregated into rural and urban segments.

The model combines national production and consumption data and solves for the optimal level of commodity supply and demand. If supply and demand relationships imply the need to trade (either import or export), prices will be those on the world market corrected for market transaction costs. When imports are implied, domestic prices equal world prices plus marketing costs; when exports are implied, domestic prices equal world prices less marketing costs. If supply and demand relationships imply no trade, then prices are determined within countries.

Regional levels of variables are aggregated from national totals. Although the model cannot specifically capture trade flows among the countries in the region, it can identify total regional demand and supply and net trade flows at the regional level, based on national exports and imports of traded commodities.

The DREAM Model

As noted in Chapter 1, a central idea in this report is that ECA countries might be missing important opportunities to benefit from regionally coordinated national agricultural development efforts. The regional multimarket model of ECA agriculture does not include sufficient detail about a range of important variables to permit examination

[7]Although the computable general equilibrium (CGE) approach is preferable for economywide and across-country analysis, there are insufficient data available to construct a highly disaggregated social accounting matrix for each country in the study. Thus it is not possible to calibrate a regional CGE model suitable for this project.

of such potential. The multimarket model crucially lacks detailed information about agricultural production technologies and the scope for agricultural technology development and dissemination. This gap precludes analysis of the effects of regionally coordinated efforts to promote productivity growth in ECA. IFPRI's DREAM model is ideally suited to that purpose.

The DREAM model allows for two impact mechanisms arising as a consequence of technical change through farm-level adoption of improved technologies or practices. The first effect, as in the multimarket model, is brought about by changes in the volume and price of commodities traded among countries. But DREAM also allows technologies themselves to spill over from one region or country to another and to be adopted in recipient regions or countries. The spillover process provides additional economic benefits (and losses) over those arising from commodity trade alone. Where they have been rigorously researched, spillover benefits have been shown to account for half, and sometimes more, of the total benefits of agricultural research (Alston 2002).

DREAM allows for spillover time lags and for differences in the use or effectiveness of a technology between the "spillout" (source) and "spillin" (recipient) regions or countries. For example, a new pest resistant variety of maize might be developed in Kenya. Adoption of this variety in Kenya might increase maize output and place downward pressure on maize prices. It might also reduce maize imports and have a negative effect on Ugandan producers. However, assuming regional institutional and regulatory processes were in place, the germplasm itself could be utilized in Uganda and other countries in the region. This scenario would involve additional lag times in the transfer process, and different levels of adaptation or different agronomic packages would be developed in different spillin locations. DREAM helps analysts examine the pattern of potential subnational, national, and regional costs and benefits of alternative technology development and deployment strategies. In particular, DREAM can help determine whether it is best to invest in domestic research programs, to strike partnerships, or to use such facilities as ASARECA's regional research networks to access new spillin technologies.

Together these analytical tools permit examination of a range of issues central to agricultural development. Ultimately, their application sheds light on such questions as: What are the implications of continuation of recent growth trends in key agricultural subsectors and in the agricultural sector as a whole? What levels of growth would be required to achieve key development targets? How do different agricultural subsectors compare for their potential effects on agricultural gross domestic product (GDP) and overall GDP? How do different subsectors compare for their effects poverty? What are the poverty-reducing effects of growth in nonagricultural sectors? Which combinations of agricultural and nonagricultural investment yield the greatest changes on overall growth and poverty? Are there any important subnational differences in subsectoral priorities? How do returns to productivity-enhancing investments compare to those that result from reductions in barriers to trade and marketing? What are the potential benefits from regional cooperation in agricultural development? These questions are addressed in Chapter 5. To set the stage for the application of the models, Chapter 3 outlines key aspects of the socioeconomic and biophysical underpinnings of agriculture in ECA, and Chapter 4 details recent trends and current conditions in the region.

CHAPTER 3

Socioeconomic and Biophysical Underpinnings

Agricultural performance both derives from and conditions socioeconomic and biophysical realities (Pender, Place, and Ehui 1999; Pender 2004). In particular, it determines and reflects spatial distributions of human population and associated access to cultivable land, agricultural potential as captured by agroecological conditions, and access to markets (Wood et al. 1999). Using outputs of the spatial analysis outlined in Chapter 2, this chapter describes these realities. A basic argument underlying the analysis is that areas exhibiting different (or similar) combinations of these characteristics are often associated with different (or similar) management practices and livelihood strategies, and thus overall agricultural performance (Nkonya et al. 2004). Much of the discussion therefore revolves around a series of mapped and tabular representations of population density, agricultural potential, market access, and development domains that identify areas endowed with similar realizations of these three attributes. The varying degree of completeness and reliability the data used and the exploratory nature of some of the spatial modeling techniques employed (see Appendix A) renders tentative some of the conclusions that can be drawn from the analysis. This is especially true for the discussion of market access. However, as pathbreaking first approximations of the phenomena under consideration, the analytical results represent major contributions to the understanding of agriculture in eastern and central Africa (ECA).

Population Distribution and Agricultural Land Use

The land:labor ratio has been theorized to have consequences for land management and production technology choice (Boserup 1981). Holding other factors constant, farmers in densely populated areas are more likely to undertake labor-intensive production strategies than are those in areas of low density. Population density is therefore a potentially useful tool for understanding fundamental opportunities and constraints facing agriculture in ECA.

Figure 3.1 shows the distribution of population in ECA. The temperate and subtropical highland areas of Burundi, Ethiopia, Kenya, Rwanda, and Uganda are the predominant high-density areas of the region (more than 100 persons per square kilometer). Smaller pockets of high population density are found in northeastern DRC, areas bordering Lake Victoria, east-central Sudan, and the northern and southern border of Tanzania. The remaining areas have relatively low population densities (fewer than 100 persons per square kilometer).

Roughly one-third (300 million hectares) of ECA's total land area is devoted to agricultural uses (Table 3.1). Roughly 80 percent of agricultural land is rangeland and pasture. The remainder is under crops, primarily annuals. The relative extent, distribution, and mix of crop-

Figure 3.1 Population density

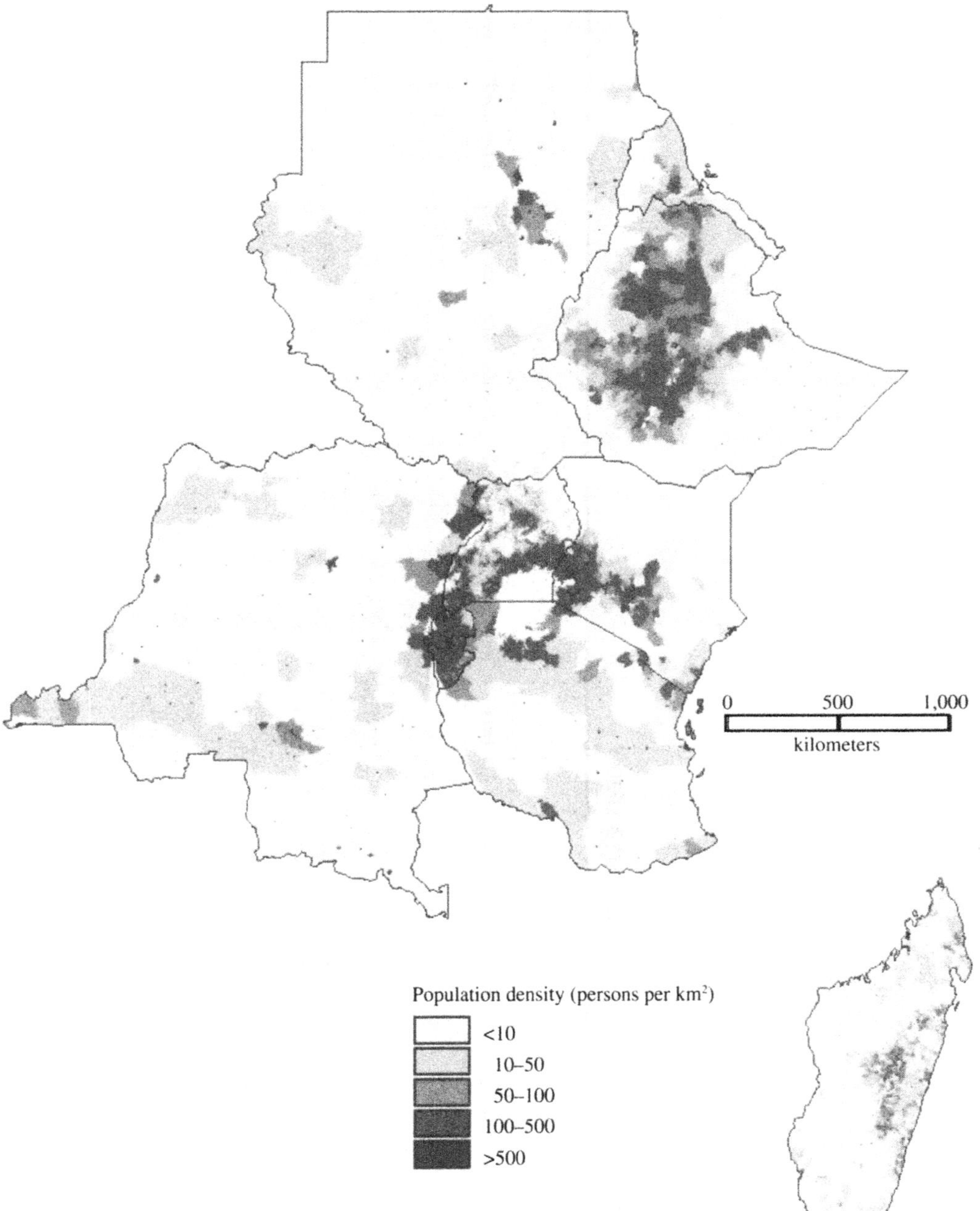

Source: Authors' calculations based on CIESIN et al. 2004.

and livestock-based agriculture vary widely across the region (Figure 3.2).[8] The agricultural area of the Democratic Republic of Congo (DRC) is slightly less than that of

[8]The map shows both the spatial intensity of cultivation or rangeland within an area by darker shading, and the relative mix of crop- and livestock-based activities according to the mix of colors: from fully green to represent almost exclusively crop-based activities (including integrated crop livestock systems) to fully brown for extensive, almost purely pastoral areas.

Table 3.1 National and ECA agricultural land uses by area and area shares

	Total land area	Agri-cultural area	Total cropland	Annual cropland	Perennial cropland	Irrigated cropland	Permanent pasture	Agri-cultural land/ total land	Cropland/ total agri-cultural	Pasture/ total agri-cultural	Annual/ cropland	Perennial/ cropland	Irrigated/ cropland
Burundi	2,568	2,170	1,351	986	365	74	990	84.5	62.3	37.7	73.0	27.0	16.8
DRC	226,705	22,800	7,800	6,700	1,100	11	15,000	10.1	34.2	65.8	85.9	14.1	4.8
Eritrea	10,100	7,470	503	500	3	21	6,967	74.0	6.7	93.3	99.4	0.6	0.0
Ethiopia	100,000	30,671	10,671	9,936	735	190	20,000	30.7	34.8	65.2	93.1	6.9	2.4
Kenya	56,914	26,462	5,162	4,600	562	90	21,300	46.5	19.5	80.5	89.1	10.9	2.1
Madagascar	58,154	27,550	3,550	2,950	600	1,090	24,000	47.4	12.9	87.1	83.1	16.9	2.2
Rwanda	2,467	1,850	1,385	1,116	269	6	465	75.0	74.9	25.1	80.6	19.4	14.5
Sudan	237,600	133,833	16,653	16,233	420	1,950	117,180	56.3	12.4	87.6	97.5	2.5	0.3
Tanzania	88,359	40,100	5,100	4,000	1,100	170	35,000	45.4	12.7	87.3	78.4	21.6	2.7
Uganda	19,710	12,312	7,200	5,100	2,100	9	5,112	65.2	58.5	41.5	70.8	29.2	17.1
ECA	802,577	305,218	59,375	52,121	7,254	3,611	246,014	38.0	19.5	80.5	87.8	12.2	2.4

Source: Compiled by authors from FAO n.d. (accessed November 2004).

Note: Areas are in thousands of hectares; ratios are percentages.

Figure 3.2 Agricultural land use

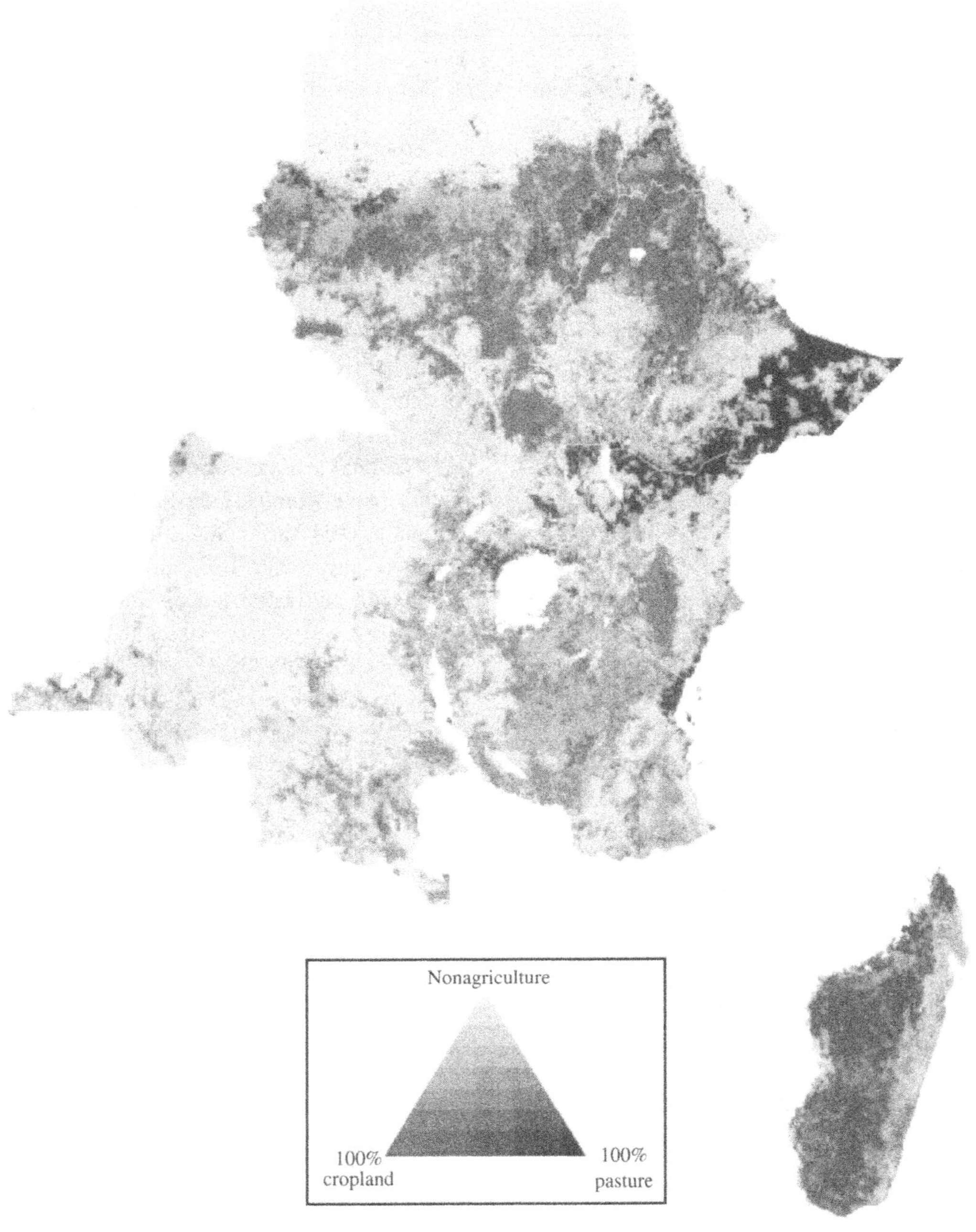

Source: Authors' calculations based on cropland and pasture data from SAGE 2004.

Kenya or Madagascar but only about one-tenth of the country's total land area. At the other extreme, in small countries, such as Burundi, Eritrea, and Rwanda, agriculture accounts for at least three-quarters of the total land area.

The distribution of agricultural production matches that of human population. The joint effects of favorable rainfall and temperature conditions and proximity to water bodies and rivers define the distribution of both people and agriculture.

In many parts of the region, crop production is only feasible under irrigated conditions. However, Sudan (with nearly 2 million hectares [ha] of irrigated area) and

Table 3.2 Per capita cultivated land by land size group

	Small	Medium	Large
Land share (%; ECA total = 100)	14.75	11.32	73.92
Rural population share (%; ECA total = 100)	68.17	10.47	21.36
Land per rural population (hectares per capita) (ECA average = 0.2)	0.04	0.21	0.69

Notes: Small = less than 80 percent of the national average; medium = between 80 and 120 percent of the national average; large = more than 120 percent of the national average.

Madagascar (with slightly more than 1 million ha) account for 84 percent of the region's irrigated area. Ethiopia and Tanzania are the only other countries with major irrigated areas (190,000 and 170,000 ha, respectively). As a whole, only 6 percent of ECA's cropland is irrigated, with Madagascar's 30 percent and Sudan's 12 percent representing the highest shares. In the more humid environments of DRC, Rwanda, and Uganda, irrigation accounts for less than 0.5 percent of cropped areas.

High concentrations of people in particular areas (Figure 3.1) suggest that access to agricultural land in ECA is constrained. For the region as a whole, cultivated land per capita stands at 0.2 ha. However, less than 11 percent of the rural population resides in areas where per capita cultivated land area is close to the regional average (Table 3.2). Almost 70 percent of the population lives in areas with less than 15 percent of the region's cultivated land; for these rural dwellers, the availability of cultivated land is only one-fifth that of the regional average.

National averages of cultivated land per capita range from 0.11 ha for Ethiopia and Kenya to 0.64 ha for Sudan (Table 3.3). In small countries like Burundi and Rwanda, per capita cultivated land is more equally distributed across size classes than in large countries like DRC and Sudan (Table 3.4). With the exceptions of Burundi and Rwanda, 60–80 percent of rural inhabitants live in areas with below-average amounts of per capita cultivable land (Table 3.5).

Agroecological Conditions and Agricultural Potential

As shown by the distribution of agricultural land use, opportunities and constraints in

Table 3.3 Per capita cultivated land by country and land size group (hectares)

	National average	Small	Medium	Large
Burundi	0.16	0.09	0.15	0.28
DRC	0.20	0.03	0.20	0.88
Eritrea	0.24	0.03	0.24	0.89
Ethiopia	0.11	0.03	0.11	0.30
Kenya	0.11	0.03	0.10	0.31
Madagascar	0.18	0.06	0.18	0.48
Rwanda	0.25	0.16	0.24	0.42
Sudan	0.64	0.10	0.64	3.28
Tanzania	0.20	0.03	0.19	0.60
Uganda	0.31	0.08	0.31	0.79

Notes: Small = less than 80 percent of the national average; medium = between 80 and 120 percent of the national average; large = more than 120 percent of the national average.

Table 3.4 Land share by size of cultivated land per rural population (%; national = 100)

	Small	Medium	Large
Burundi	24.5	29.8	45.7
DRC	13.5	5.4	81.1
Eritrea	9.9	4.7	85.4
Ethiopia	15.4	8.3	76.3
Kenya	17.9	13.2	68.9
Madagascar	19.2	15.0	65.8
Rwanda	21.3	42.1	36.7
Sudan	12.3	5.3	82.4
Tanzania	11.3	9.9	78.9
Uganda	15.5	15.7	68.8

Notes: Small = less than 80 percent of the national average; medium = between 80 and 120 percent of the national average; large = more than 120 percent of the national average.

Table 3.5 Rural population distribution by per capita cultivated land size (%; national = 100)

	Small	Medium	Large
Burundi	44.1	30.4	25.5
DRC	76.4	5.4	18.1
Eritrea	71.9	4.8	23.3
Ethiopia	64.3	8.4	27.3
Kenya	62.5	13.5	24.0
Madagascar	59.8	15.2	25.0
Rwanda	34.1	43.7	22.2
Sudan	78.5	5.3	16.2
Tanzania	64.2	9.9	25.9
Uganda	57.1	15.8	27.0

Notes: Small = less than 80 percent of the national average; medium = between 80 and 120 percent of the national average; large = more than 120 percent of the national average.

agricultural production vary by location and type of production systems (Wood et al. 1999; Nkonya et al. 2004). Within ECA, where agriculture is dominated by subsistence-oriented smallholders, the three most binding constraints influencing agricultural production potential are the availability and variability of water supply, soil fertility, and the biotic pressure from pests and diseases. In theory, all these attributes should be reflected in any measure of agricultural potential. In practice, paucity of appropriate data at appropriate scales renders such treatment infeasible. The availability of water—from rainfall, local groundwater, surface water, or formal irrigation schemes—is generally the most binding of constraints. Figure 3.3 shows the distribution of the length of growing period (LGP) across the ECA region. LGP measures the total length of time (shown in the figure in months) that rainfall exceeds evapotranspiration, leaving sufficient excess water to support the growth of crops and pasture.

Across ECA, 68 percent of cropland and 76 percent of the population fall within areas where the LGP exceeds six months per year (Table 3.6). There is considerable variation across countries. For example all of Eritrea's cropland and rural population are located in semi-arid areas with less than a four-month growing period, whereas Rwanda lies entirely within a humid zone with a growing season of eight months or more.

Recognizing that agricultural potential relies on more than rainfall alone, and on the basis of technical consultation with specialists based in the region, additional layers of information were added to generate a more complete picture of agricultural potential. In addition to LGP, data were integrated to reflect the availability of irrigation or likely access to surface water sources and the ability of the soil to support agriculture (specifically, average measures of organic matter, pH, and texture in the top soil layer and top-soil depth). Finally, for the purposes of considering future agricultural options in the region, areas situated within protected areas or at very high elevations were omitted from consideration, as were very remote (for example, large areas of DRC) or otherwise extremely marginal lands (for example, the arid lands of Sudan).

Access to Markets

To fully understand how a location's absolute agricultural potential translates into

Figure 3.3 Agricultural potential: Length of growing period

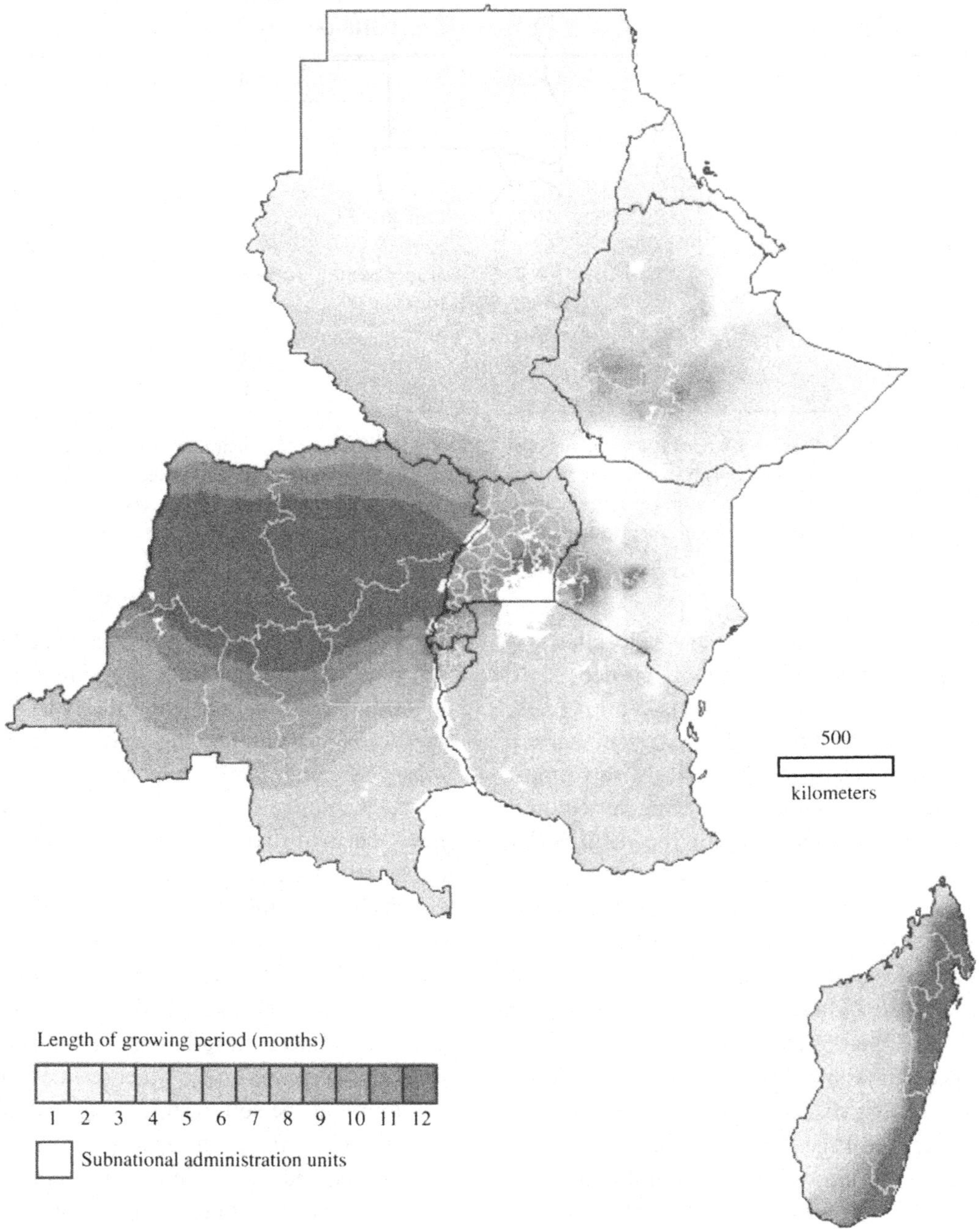

Source: Authors' calculations based on FAO/IIASA Global AEZ 2001.

comparative advantage for different production activities requires information on access to markets (Omamo 1998a,b). Opportunities for gathering market information, obtaining credit, buying inputs, and selling outputs depend on a wide range of socioeconomic, institutional, and cultural factors. Unfortunately, data describing such conditions in ECA are patchy and of questionable quality. Constructing a market-access spatial layer therefore resides firmly in the domain of modeling and spatial extrapolation. The appendixes detail how data on some of the abovementioned factors are combined and extrapolated to yield a picture of market access in ECA that, although incomplete, is a useful entry point for this crucial determinant of agricultural opportunities and constraints.

This study focuses on a set of criteria that reflect the physical accessibility of a range of markets in terms of expected travel times.[9] Five distinct types of market opportunity are identified: localized trade/exchange, subregional trading centers, central urban markets, transborder trade (by road, rail, or water), and international fresh markets accessed by airports. For each type of market, individual measures of market access are generated, spanning the whole region. Information on road location and quality, slope, and off-road land cover is combined to assess travel times to target market locations. Figure 3.4a shows the results for one type of market—subregional market centers—that are defined as having a population of more than 50,000 people. Significant areas in northern Sudan, eastern Ethiopia, northern Kenya, and central DRC are very remote from regional trading centers. For the region as a whole, more than 40 percent of the rural population and cropland areas are more than eight hours' travel away from such markets, and only 14 percent are within two hours of travel time (Table 3.7). In Burundi, Rwanda, and Uganda, slightly more than half the population is within four hours of travel from a regional market. Conversely, in DRC and Tanzania, more than half the rural population is more than eight hours' travel away.

A similar analysis is performed for central urban, transborder, and high-value airfreight markets, using as market targets capital cities (and other major cities, such as Mombasa in Kenya), border crossing points, and international airports, respectively. In all these cases, areas within three hours' travel of target markets are classified as having high access levels. In the case of local trade or exchange, a different measure is used. The opportunity for local trade or exchange is considered high for any location for which 300 or more people live within a 5-kilometer radius.

To assess the overall level of market access of any location, the numbers of different types of market to which that location has high levels of access are added together. Figure 3.4b shows locations having high levels of access to two or more types of market. Areas proximate to the major trade corridors show up as high access, as do areas surrounding capitals in the high-density highlands of Burundi, Ethiopia, and Rwanda. Elsewhere, high-access areas are more restricted; they constitute areas contiguous with urban centers.

Agricultural Development Domains

Figure 3.5 illustrates the intersection of the three socioeconomic and biophysical layers strongly related to the feasibility and attractiveness of specific development and livelihood strategies: population density, agricultural potential, and market access. The distinct areas delineated on this map are defined as agricultural development domains—areas for which a given agricultural development strategy is likely to have similar relevance (Wood et al. 1999). Development domains are defined using consistent data and criteria across the region, thus helping diagnose development constraints and formulate and evaluate strategic intervention options in comparable ways.

To facilitate analysis and communication, only eight domain types are defined. This breakdown is done by classifying each of the three key factors into two values: high or low. Population densities are assumed to be high at densities of 100 persons per

[9]Off-road travel time is assumed to be by foot, with walking speed conditioned by slope and land cover. On-road travel time is assumed to be by motorized vehicle, with road speed conditioned by road quality and slope.

Table 3.6 National and ECA distribution of cropland and rural population by length of growing period category

Length of growing period (months)	Burundi		Eritrea		Ethiopia		Kenya		Madagascar	
	Rural population	Cropland	Rural population	Cropland	Rural population	Cropland	Rural population	Cropland	Rural population	Cropland
<4			100	100	13	10	13	31	5	9
4–6					12	18	18	30	12	15
6–8	16	28			37	39	11	8	29	12
>8	84	72			38	33	58	31	54	64
Total	100	100	100	100	100	100	100	100	100	100

square kilometer or greater and low otherwise; agricultural potential is assumed to be high where LGP is 180 days or more and low otherwise; and market access is assumed to be high in locations with high levels of access to at least two of the five types of market and low otherwise. Domains are described by their high or low status in the sequence—agricultural potential, market access, and population density, as shown in the legend to Figure 3.5. For instance, HHL denotes high agricultural potential, high market access, and low population density.

Despite the limited number of domains, the spatial variability of domains can be

Figure 3.4 Agricultural market access: Travel times to regional market centers and aggregate high/low market access classes

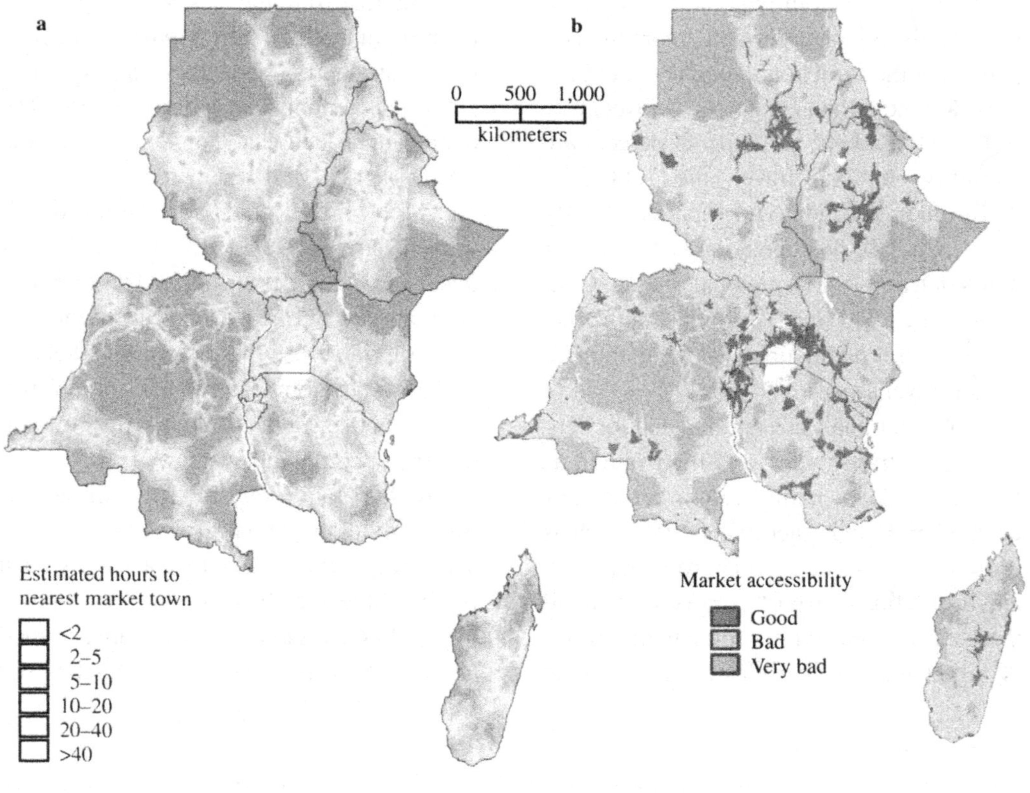

Source: Authors' calculations.

Rwanda		Sudan		Tanzania		Uganda		DRC			
Rural population	Cropland	Rural population	Cropland	Rural population	Cropland	Rural population	Cropland	Rural population	Cropland	ECA rural population	ECA cropland
		66	57	1	1					14	15
		19	35	28	25		3		3	11	19
		14	7	64	70		4	10	46	24	32
100	100	2	1	7	4	99	93	90	51	52	34
100	100	100	100	100	100	100	100	100	100	100	100

quite complex, especially in highland areas, reflecting marked local changes in agricultural potential, market access, or population density. Domains straddle national and subnational boundaries where development conditions are similar.

Table 3.8 summarizes the distribution of some key measures within the eight domain types.[10] The largest individual domain is HLL (38 percent of ECA land area) followed by LLL (20 percent). Areas with high agricultural potential and high market access account for only 4.4 percent of the land area but include more than 11 percent of cropland and 19 percent of the rural population. The proportion of cropland to total land area falls markedly as areas become less suitable. Domain HHH has 2 percent of total land and more than 7 percent of cropland, HLL has 38 percent of land area and about the same percentage of crop area, whereas LLL has 21 percent of land area and 16 percent of cropland. More than 60 percent of the rural population and almost 60 percent of the cropland can be found in the 45 percent of ECA area with high potential. But more than 40 percent of the population and almost 50 percent of the cropland are located in areas with low market access. Country-specific breakdowns are shown in Appendix A. Most countries contain at least six of the domain types.

As noted in Chapter 2, the development domain approach allows spatially disaggregated analysis of alternative development strategies. Table 3.9 links each of the eight development domains to specific development strategies, and gives examples of where in the ECA region each domain occurs. Even in situations with lowest agricultural development potential (LLL), there are multiple development options, some of which are complementary. A given strategic approach—for example, promotion of high-input cereals—might be relevant to several domains, but detailed implementation may differ across domains—for example, because of differences in dominant crop mixes, or in degrees of crop-livestock interactions. These principles are discussed at greater length in the appendixes.[11]

The analysis in Chapter 5 builds understanding about specific agricultural development options, focusing on their implications for growth and poverty reduction. In Chapter 6, the scheme in Table 3.9 is shown

[10]The "not included" category refers to protected and very remote areas not captured in the classification scheme.

[11]A very specific and important example of this need to further disaggregate domains and strategies is illustrated in Figure 3.5. In that figure, the 1200-meter contour broadly accepted as a useful means of distinguishing between highland and lowland production conditions is superimposed. As specific development options are articulated and the process of strategy formulation and design proceeds, it will be important to zoom in—both geographically and by production systems—to better articulate operational intervention options.

Table 3.7 National and ECA distribution of cropland and rural population by market access zones (%)

Access to towns >50,000 (hours)	Burundi		Eritrea		Ethiopia		Kenya		Madagascar	
	Rural population	Cropland	Rural population	Cropland	Rural population	Cropland	Rural population	Cropland	Rural population	Cropland
<2	58	49	38	9	18	16	37	17	18	8
2–4	34	39	29	29	31	26	38	25	24	19
4–6	8	8	15	34	24	29	19	25	18	18
6–8	1	3	10	20	13	17	6	16	15	20

to provide a potent organizing framework for linking the potentially rewarding strategic directions (identified in Chapter 5) to specific opportunities and constraints in ECA agriculture, as illuminated by the eight development domains. But first Chapter 4 summarizes recent trends and current performance in ECA agriculture.

Figure 3.5 Agricultural development domains and administrative boundaries

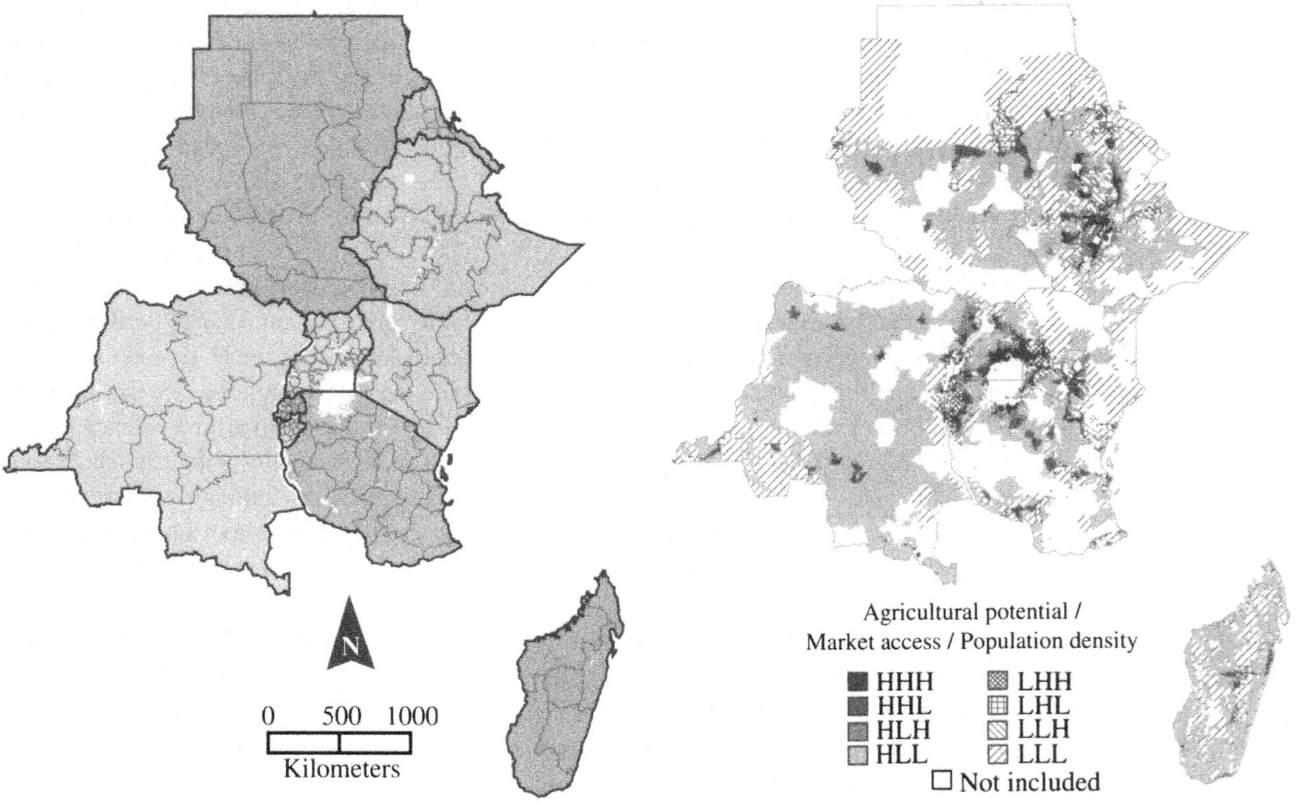

Source: Authors' calculations.

Note: "H" and "L" refer to the following characteristics: agricultural potential, market access, and population density, in that order. See inside back cover of this report for a color version of this map.

Rwanda		Sudan		Tanzania		Uganda		DRC			
Rural population	Cropland	Rural population	Cropland	Rural population	Cropland	Rural population	Cropland	Rural population	Cropland	ECA rural population	ECA cropland
54	47	22	16	20	15	47	32	12	6	25	16
33	34	28	28	30	27	32	36	17	12	28	25
7	9	19	26	22	22	13	29	17	13	19	21
4	5	12	16	13	14	4	5	13	11	11	14

Table 3.8 Distribution of populations, lands, and cattle by ECA agricultural development domain

Domain	Total population	Rural population	Land area (hectares)	Crop area (hectare)	Pasture area (hectares)	Cattle (herd)
HHH	48,426,587	29,484,613	15,455,083	4,841,643	5,597,020	8,247,287
HHL	17,401,314	9,748,804	20,422,786	2,603,418	7,124,266	5,505,638
LHL	36,460,016	31,859,498	21,068,213	5,953,171	7,282,505	11,922,444
HLL	67,321,092	58,457,657	313,269,857	25,809,309	85,381,313	41,148,856
LHH	28,541,396	14,476,029	6,338,972	1,879,020	2,339,045	3,305,366
LHL	11,367,976	5,816,946	10,867,703	1,308,882	4,116,101	3,241,843
LLH	15,142,158	13,197,227	8,299,234	2,288,213	3,111,631	4,420,335
LLL	35,420,112	31,374,157	170,617,605	10,386,762	53,824,743	22,289,954
Not included	21,559,349	13,125,585	258,650,848	11,815,719	48,959,981	15,794,770
Total	281,640,000	207,540,516	824,990,328	66,886,138	217,736,605	115,876,493
			(percent)			
HHH	17.2	14.2	1.9	7.2	2.6	7.1
HHL	6.2	4.7	2.5	3.9	3.3	4.8
HLH	12.9	15.4	2.6	8.9	3.3	10.3
HLL	23.9	28.2	18.0	38.6	39.2	35.5
LHH	10.1	7.0	0.8	2.8	1.1	2.9
LHL	4.0	2.8	1.3	2.0	1.9	2.8
LLH	5.4	6.4	1.0	3.4	1.4	3.8
LLL	12.6	15.1	20.7	15.5	24.7	19.2
Not included	7.7	6.3	31.4	17.7	22.5	13.6
Total	100	100	100	100	100	100

Note: There are eight development domains defined by three factors (agricultural potential, market access, and population density) that each take on one of two values (high or low). H = high; L = low.

Table 3.9 Agricultural development options within ECA agricultural development domains

Agricultural potential	Market access	Attributes	Example locations in ECA and potential agricultural development/livelihood options — Population density: High	Low
High	High	Greatest commercialization and diversification options	*Example locations:* Parts of central and western Kenya, Uganda's Lake Victoria Crescent, parts of central and southwestern and southeastern highlands of Ethiopia, parts of Rwanda and Burundi *Options:* High-input cereals (for example, maize, rice, wheat) Perishable cash crops (for example, vegetables, fruit, flowers, ornamentals) Intensive livestock (for example, dairy, chicken, pig) Nonperishable cash crops (for example, coffee)	*Example locations:* Isolated areas scattered throughout region *Options:* As for high population density plus more extensive high-value options Cash crops: cotton, tea, oil crops Perennials: fruit trees
High	Low	More limited technology adoption and commercialization	*Example locations:* Southwestern Uganda, parts of central and western Kenya, much of the Ethiopian highlands, northern Tanzania, Rwanda, and Burundi *Options:* High-input cereals (for example, maize, rice, wheat) Nonperishable cash crops	*Example locations:* Large areas of all countries: most of central DRC, southern Sudan, parts of central Uganda, Kenya, and Tanzania, widely scattered areas in Ethiopia and Madagascar *Options:* High-input cereals Nonperishable cash crops Livestock intensification; improved grazing areas
Low	High	Commercialization options for high-input, labor intensive production	*Example locations:* Parts of northern Ethiopia and central Eritrea, north-central Sudan, western Kenya, Rwanda, and Burundi *Options:* With irrigation investment High-input cereals Perishable cash crops Dairy, intensive livestock Without irrigation investment Low-input cereals	*Example locations:* Isolated areas scattered throughout region *Options:* With irrigation investment High-input cereals Perishable cash crops Dairy, intensive livestock Without irrigation investment Low-input cereals Livestock intensification; improved grazing areas Woodlots
Low	Low	Little comparative advantage in intensive production and high levels of input use	*Example locations:* Northern and eastern highlands of Ethiopia, parts of Western Kenya, Rwanda, eastern DRC near lakes Edward and Kivu *Options:* Low-input cereals Limited livestock intensification Emigration	*Example locations:* Some lowland areas in Ethiopia and Eritrea, central Sudan, southeastern Kenya, eastern DRC *Options:* Low-input cereals Livestock intensification; improved grazing areas

Sources: Compiled by authors drawing extensively on empirical research in ECA, especially in Uganda, Ethiopia, and Kenya (in particular, see Pender, Place, and Ehui 1999; Pender 2004; Ehui and Pender 2005).

CHAPTER 4

Recent Trends and Current Performance

It is well known that agriculture is the most common occupation in eastern and central Africa (ECA). In addition to its ubiquity across the region, agriculture in ECA is also diverse. Chapter 3 demonstrated that this heterogeneity stems in large part from varying underlying socioeconomic and biophysical conditions. A basic premise of this study is that ECA's agricultural diversity can also be traced to circumstances and policies that affect the degree to which agricultural economies are integrated into national economies and regional and international markets. Agricultural systems range from those based primarily on annual staples destined for home consumption to systems in which perennial cash crops are prominent. In some systems, production depends almost entirely on domestic labor; in others purchased inputs are more important. However, one feature that the agricultural sectors of most ECA countries do have in common is that none of them would appear to be performing at levels required to contribute meaningfully to growth, poverty reduction, and overall economic and social development.

In this chapter, key aspects of that disappointing performance are highlighted. The aim is not to dissect the minutiae of each and every indicator of agricultural performance in ECA but rather to set the stage for the exploration (in Chapter 5) of alternative scenarios for agricultural development in the region. First, the place of agriculture in the regional economy is outlined, focusing on the varying shares of agriculture in overall gross domestic product (GDP) and on recent subsectoral and sectoral growth rates. Recent trends and current levels of agricultural productivity, production, consumption, and trade in ECA are then described in detail. A decidedly gloomy picture emerges.

GDP and Agricultural GDP

Agriculture looms large in national economies throughout ECA (Figure 4.1). Agriculture accounts for 43 percent of total GDP in the region. In five countries (Burundi, DRC, Ethiopia, Sudan, and Tanzania), agriculture's share of GDP exceeds 50 percent. Only in Eritrea, Kenya, and Madagascar does it contribute less than 30 percent to GDP.[12] The distribution of the region's GDP closely matches that of the region's agricultural GDP (AgGDP; Figure 4.2). Countries with relatively large (small) national economies also have relatively large (small) agricultural economies. The largest economies are those of Kenya, Tanzania, Uganda, the DRC, Ethiopia, and Madagascar; those having the smallest are Eritrea, Burundi, and Sudan.[13] Whereas per

[12]Note, however, that the reasons for these relatively low shares are different for the three countries. Eritrea has little agricultural land; Kenya's structural transformation toward a less agriculture-based economy is more advanced than in other countries in the region; Madagascar's large agricultural potential remains mostly untapped.

[13]With a population of 32 million people, oil reserves, and a high potential for irrigated agriculture, Sudan may not reside in the small-economy category for very long.

Figure 4.1 Agriculture's share of GDP

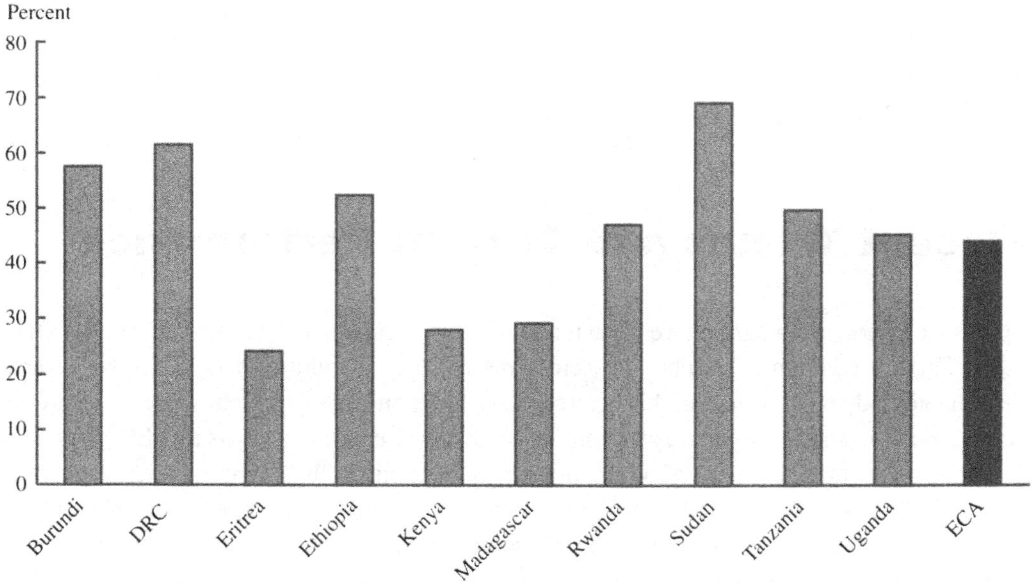

Sources: FAO n.d. and World Bank development indicators.

capita overall GDP levels range from about US$400 (Kenya) to slightly over US$100 (Ethiopia), per capita AgGDP levels in most countries in the region cluster around US$100, with some (Eritrea, Ethiopia, Burundi, and DRC) hovering close to US$50 (Figure 4.2).

Agricultural Productivity and Growth

Figure 4.3 plots land and labor productivity in agriculture between 1961 and 2000 for ECA and other regions. The land productivity measure is the ratio of gross output to the total hectares used in agriculture, whether

Figure 4.2 Per capita GDP and agricultural GDP, 2003

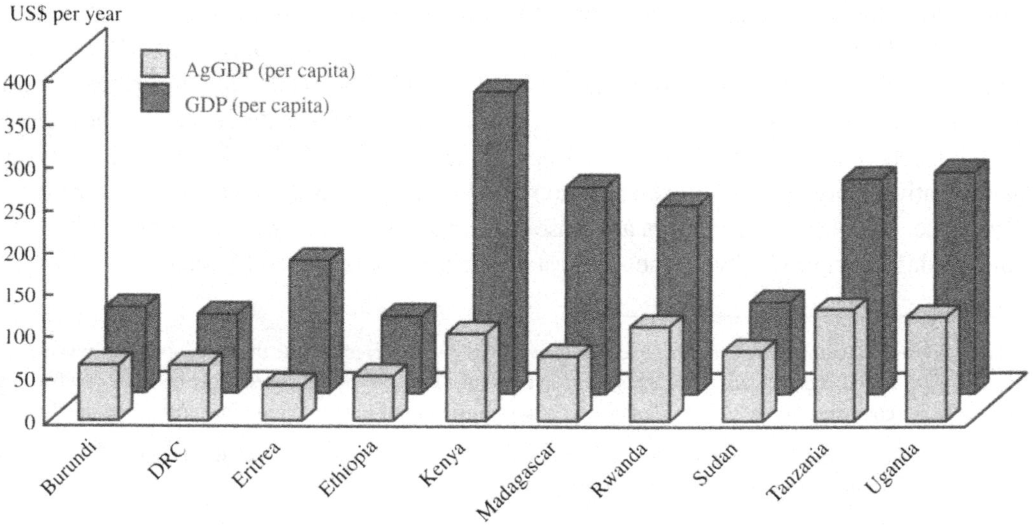

Sources: FAO n.d. and World Bank development indicators.

Figure 4.3 Agricultural land and labor productivity, 1961–2000

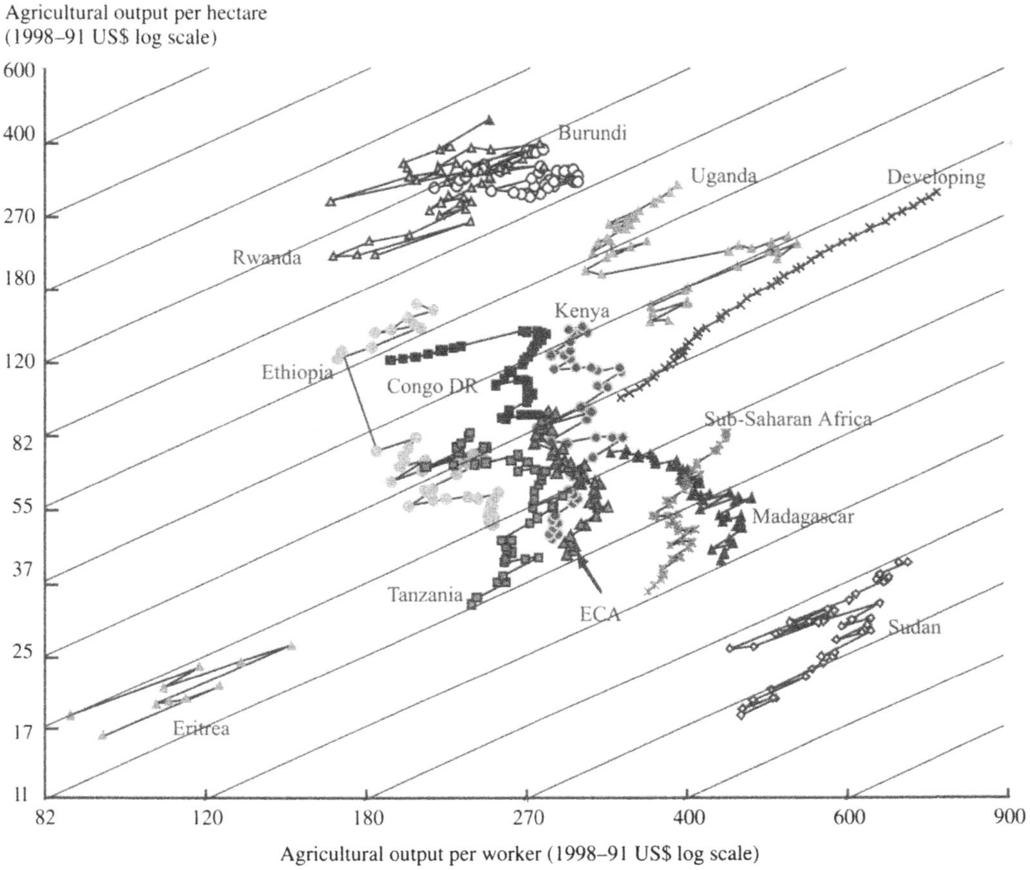

Source: Authors' calculations based on FAO n.d. (accessed 2003).

irrigated or nonirrigated cropland, pastureland, or rangeland. The labor productivity measure captures gross output relative to the economically active agricultural population. The diagonal lines indicate constant factor ratios. When a country or region's productivity locus is flatter (steeper) than these diagonal lines, it indicates an increase (decrease) in the number of hectares per worker over time. The locus for Africa as a whole is steeper than the diagonals, indicating declines in labor productivity. Labor productivity in ECA has declined substantially. The contraction has been so marked that the region actually produced less per worker in 2000 than it did four decades earlier. This regionwide contraction in labor productivity has of course been based on contractions in several ECA countries, most notably in DRC, Kenya, Madagascar, and Tanzania. Labor productivity in Ethiopia, Rwanda, Sudan, and Uganda has recovered substantially in recent years.

Given these trends in agricultural productivity in ECA, it is not surprising that average yields for ECA's major crops currently fall well below those elsewhere in Africa, and even further below global levels (Table 4.1). Only for cassava, beans, coffee, and tea do ECA yields compare favorably with average African and global levels.

These trends in productivity growth have translated into poor overall agricultural growth rates in individual ECA countries and for the region as a whole (Table 4.2). Overall performance for ECA agriculture in 1993–2003 was slightly better than in the preceding decade. But at 2.05 percent, agricultural growth did not keep pace with population growth, which stood at close

Table 4.1 Agricultural commodity yields: ECA, Africa, and global, 2003

Commodity	ECA	Africa	Global
Maize	1.39	1.61	4.47
Rice	1.12	1.87	3.84
Wheat	1.38	2.03	2.66
Sorghum	0.67	0.88	1.30
Millet	0.47	0.70	0.82
Potatoes	7.46	11.17	16.45
Sweet potatoes	4.29	4.32	13.49
Cassava	8.18	8.83	10.76
Beans	0.60	0.62	0.70
Groundnuts	0.62	0.86	1.35
Sugarcane	4.11	56.75	65.29
Bananas	4.69	6.59	15.25
Coffee	0.57	0.45	0.75
Tea	1.85	1.96	1.33
Barley	1.18	1.24	2.48
Oilseeds	0.51	0.69	1.75
Beef (kg/animal)	127.	148.	200.
Chicken (kg/animal)	0.92	1.17	1.72
Cow milk (kg/animal/year)	427.	496.	2,197.

Source: FAO n.d. (accessed 2004).
Notes: Units are metric tons per hectare unless otherwise indicated.

Table 4.2 Agricultural growth rates in ECA countries, Africa, and developing regions, 1983–2003 (%)

	Growth rates for crops					
	1983–93			1993–2003		
	Production	Area	Yield	Production	Area	Yield
Burundi	2.73	1.42	1.30	−0.09	0.31	−0.40
DRC	2.87	2.83	0.04	−2.51	−2.11	−0.41
Eritrea	na	na	na	−1.37	0.86	−2.21
Ethiopia	1.16	−0.84	2.02	4.06	3.14	0.89
Kenya	4.17	3.23	0.90	1.87	0.45	1.41
Madagascar	1.51	0.61	0.90	0.66	0.45	0.21
Rwanda	0.01	2.50	−2.43	7.86	6.48	1.30
Sudan	1.47	0.82	0.65	3.98	1.46	2.48
Tanzania	0.47	1.21	−0.74	1.73	0.47	1.25
Uganda	3.10	2.73	0.36	3.31	2.31	0.98
ECA	2.07	1.45	0.62	2.05	1.19	0.84
Developing countries	2.86	1.05	1.79	3.00	0.74	2.24
Sub-Saharan Africa	3.90	2.93	0.94	2.73	1.84	0.88
World	1.83	0.38	1.44	2.24	0.27	1.96

Source: FAO n.d. (accessed May 2006).
Note: na indicates not available.

to 3 percent over this period. Only countries with low agricultural growth rates initially—that is, those emerging from civil strife—registered growth rates high enough to offset population growth.

Agricultural Production, Consumption, and Trade

The bleak picture for agricultural productivity growth in ECA has major implications for aggregate relationships among agricultural production, consumption, and trade in the region. Across the region, consumption exceeds production for several commodities (Table 4.3). Indeed, most countries in ECA are net importers of most agricultural commodities (Table 4.4). Only for coffee, tea, and fruits and vegetables does production exceed consumption consistently, leading to regular exports. Kenya, Tanzania, Uganda, and Ethiopia account for the bulk of those exports. Kenya is also the region's principal importer of agricultural commodities.

Between 1996 and 2000, the annual value of ECA agricultural trade amounted to slightly more than US$20 billion per year (Table 4.5). Total exports to non-ECA countries yielded US$4 billion per year, with traditional exports accounting for 60 percent of that total. Cross-border trade in all commodities within the ECA region yielded

Table 4.3 Supply, demand, and net supply of selected agricultural commodities, 2003 (000s tons)

Commodity	Supply	Demand	Net supply
Maize	10,546	10,803	–257
Rice	2,558	3,069	–511
Wheat	2,015	5,026	–3,011
Sorghum	5,270	5,321	–51
Barley	1,060	1,092	–32
Millet	1,701	1,702	–1
Oats	55	55	0
Other cereal	1,690	1,681	9
Potatoes	3,137	3,181	–44
Sweet potatoes	6,426	6,426	0
Cassava	30,387	30,386	1
Other roots	4,483	4,481	2
Beans	1,359	1,330	29
Peas	1,339	1,416	–77
Groundnuts	1,474	1,459	15
Sesame seed	336	197	139
Other oil crops	2,052	2,034	18
Vegetable oil	330	961	–631
Raw sugar	1,581	1,887	–306
Vegetables	9,844	8,666	1,178
Bananas	15,335	15,334	1
Fruits	3,940	3,789	151
Coffee	681	208	473
Tea	351	52	299
Spices and beverages	6,326	6,326	0
Cotton lint	168	86	82
Beef	1,290	1,294	–4
Mutton	432	425	7
Poultry eggs	468	484	–16
Other meat	573	574	–1
Milk	9,089	9,266	–177
Fish	1,232	1,230	2

Table 4.4 Net imports of major traded agricultural commodities, 1998–2001 average (000s metric tons)

	Coffee	Tea	Fruits and vegetables	Maize	Rice	Wheat	Potatoes	Vegetable oils	Sugar
Burundi	–26.70	–6.44	2.49	6.69	3.01	11.40	0.06	1.29	–0.52
DRC	0.00	0.00	14.78	63.54	57.77	297.61	39.17	25.68	51.96
Eritrea	0.00	0.27	7.05	2.79	2.33	212.51	0.00	9.93	9.16
Ethiopia	–0.17	–105.73	–6.37	24.80	5.35	831.10	–2.81	74.86	0.08
Kenya	–64.30	–233.71	–182.20	281.01	92.88	578.00	–0.21	479.37	145.75
Madagascar	–12.32	–0.30	–20.13	2.39	130.47	94.18	0.20	42.57	32.05
Rwanda	–12.42	–14.14	2.24	19.82	16.01	13.91	0.32	21.55	15.45
Sudan	9.40	20.20	5.21	39.46	31.11	676.44	0.32	114.28	–105.53
Tanzania	–46.82	–22.23	–7.27	115.93	138.58	216.01	7.48	223.01	119.75
Uganda	–186.23	–22.48	0.11	–1.88	41.06	80.07	0.12	137.31	52.30
ECA	–339.55	–384.56	–184.10	554.56	518.57	3,011.22	44.65	1,129.86	320.44

Source: FAO n.d.
Note: Negative numbers are exports.

just US$300 million. Nontraditional exports accounted for almost 43 percent of this amount, traditional exports for about 29 percent, staples roughly 20 percent, and other items about 9 percent. Domestic (within-country) trade in food staples generated almost US$16 billion—more than 50 times the value of cross-border trade within the region.

ECA's share of global agricultural exports has declined in recent years, from a high of 14 percent in the mid-1980s to less than 6 percent by 2002 (Figure 4.4). Given the prominence of coffee in ECA's agricultural exports, this overall decline has closely matched the region's falling share of the global coffee market. That share has tracked the downward trajectory of global coffee prices.

Poverty, Hunger, and Malnutrition

Because most of the region's population resides in rural areas and depends on agriculture for income and sustenance, and given the low levels of productivity growth in the sector, poverty rates are high and hunger and malnutrition have deepened in recent years. Poverty rates vary across countries but in general are very high (Table 4.6). Between 1979 and 2000, the number of

Table 4.5 Value, destination, and composition of agricultural trade, 1996–2000 average

Traded items and destinations	Value (US$ billions)	Share (%)
Traditional exports to non-ECA countries	2.4	11.9
Nontraditional exports to non-ECA countries	1.3	6.4
Other exports to non-ECA countries	0.3	1.5
Cross-border trade within ECA	0.3	1.5
Domestic markets for food staples	15.9	78.7
Total	20.2	100.0

Notes: All figures are 1996–2000 averages except for domestic markets, which are 2000 figures. Domestic market demand includes the value of own consumption. Traditional exports = coffee, tea, cotton, tobacco, cashew nuts, sugar, other fibers, cocoa, and other nuts; nontraditional exports = fish, vegetables and fruits, oilseeds, oils and fats, and processed food; staples = maize, cassava, other cereals, beans, and livestock products.

Figure 4.4 ECA shares of global agricultural and coffee exports, 1980–2002

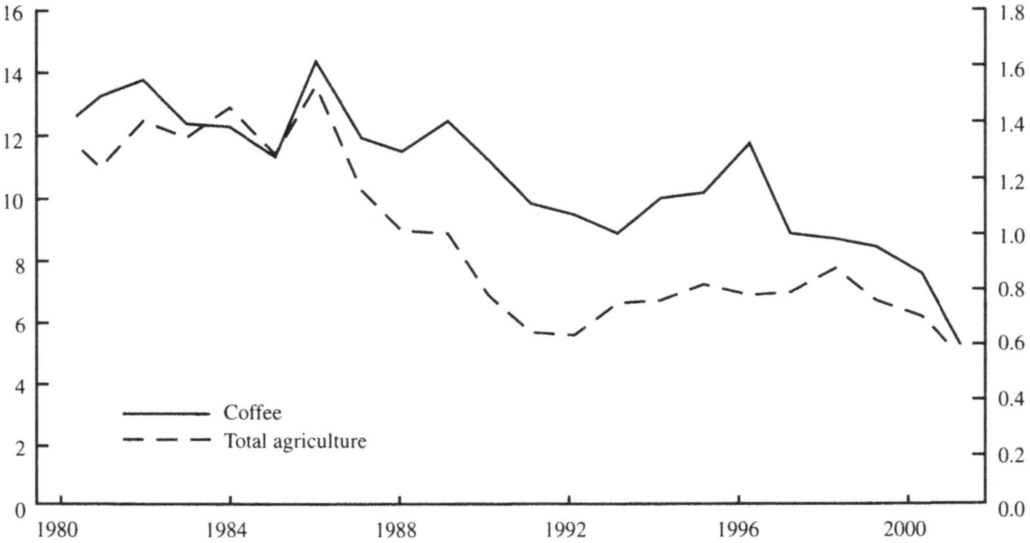

Source: FAO n.d.

malnourished adults in ECA grew faster than the overall population (Table 4.7). Rates of child undernourishment and child mortality—which is closely linked to malnutrition—stood above those for Sub-Saharan Africa and other developing regions of the world (Table 4.8).

Summary

The picture that emerges for ECA from this overview is one of a region of countries progressively less able to meet the needs of their burgeoning populations. With agriculture looming so large in most national economies, sluggish growth in agricultural

Table 4.6 Poverty rates, 2003

Country	US$1/day[a]	US$2/day[b]	National[c]
Burundi	58.4	89.2	na
DRC	na	na	na
Ethiopia	26.3	80.7	44.2
Eritrea	na	na	53.0
Kenya	22.8	58.3	42.0
Madagascar	61.0	85.1	71.3
Rwanda	51.7	83.7	51.2
Sudan	na	na	na
Tanzania	19.9	59.7	35.7
Uganda	85.0	na	55.0

Source: United Nations 2005, Table 3.
Note: na indicates data not available.
[a]Percentage of population living on less than US$1 per day.
[b]Percentage of population living on less than US$2 per day.
[c]Percentage of population living below the national poverty line (specific for each country).

Table 4.7 Adult malnutrition, 1979–2000

	Total population			Growth rate[a] (%)	Number of undernourished adults			Growth rate[a] (%)	Proportion undernourished (%)			Growth rate[a] (%)
	1979–81	1990–92	1998–2000	1980–99	1979–81	1990–92	1998–2000	1980–99	1979–81	1990–92	1998–2000	1980–99
ECA	118.7	164.6	204.1	2.9	36.7	59.6	96.3	5.0	30.9	36.2	47.2	2.1
Burundi	4.1	5.7	6.3	2.3	1.6	2.8	4.3	5.2	39.0	49.1	68.3	2.9
DRC	26.9	38.5	49.6	3.2	8.9	12.3	36.4	7.1	33.1	31.9	73.4	3.9
Kenya	16.4	24.3	30	3.2	4.0	11.5	13.2	6.5	24.4	47.3	44.0	3.3
Madagascar	9.1	12.3	15.5	2.8	1.8	4.3	6.2	6.6	19.8	35.0	40.0	3.8
Rwanda	5.2	6.4	7.0	1.6	1.2	2.2	2.8	4.5	23.1	34.4	40.0	2.9
Sudan	19.3	25.4	30.4	2.4	5.6	7.8	6.5	0.9	29.0	30.7	21.4	–1.5
Uganda	12.5	17.8	22.6	3.1	4.1	4.1	4.7	0.7	32.8	23.0	20.8	–2.4
Tanzania	18.8	27	34.3	3.2	5.2	9.8	16.2	6.0	27.7	36.3	47.2	2.8
Eritrea	na	na	3.5	na	na	na	2.0	na	na	na	57.1	na
Ethiopia	na	na	61.4	na	na	na	27.1	na	na	na	44.1	na
Sub-Saharan Africa	343.8	474.5	587.5	2.8	125.4	166.4	195.9	2.4	36.5	35.1	33.3	0.5

Source: FAO 2002.

Note: na indicates data not available.

[a]Growth rates are least squares estimates between the three periods 1979–81, 1990–92, and 1998–2000.

Table 4.8 Child malnutrition and mortality, 1990–2000

Country	Children undernourished[a] (%)	Mortality rate (per 1,000 births)
Burundi	45	190
DRC	34	205
Eritrea	44	97
Ethiopia	47	176
Kenya	23	120
Madagascar	33	137
Rwanda	29	203
Sudan	17	108
Uganda	26	145
Tanzania	29	165
Sub-Saharan Africa	29	178
Developing countries	28	91

Source: UNICEF 2001.
Note: Data are for children younger than five years old.
[a]Defined as those who are underweight, stunted, and/or wasted.

productivity has translated into sluggish overall growth and generally low income levels. High levels of agricultural importation—particularly of staples—would appear to be only partially filling the consumption needs of a population lacking purchasing power, resulting in extensive adult and child malnutrition and towering child mortality rates.

CHAPTER 5

Alternative Futures

To understand the strategic opportunities for agricultural development in eastern and central Africa (ECA), this chapter considers the implications for overall economic growth and poverty reduction of alternative scenarios for agricultural growth. A central piece of the analysis is a business-as-usual outcome that projects recent trends in agricultural growth. The business-as-usual scenario serves as a base against which to evaluate alternative agricultural development strategies for ECA.

Business-as-Usual

As illustrated in the previous chapter, central descriptors and determinants of the challenges currently facing ECA agriculture are the low growth rates in key agricultural subsectors. Consider growth rates for three agricultural commodity groups: staples, cash crops, and livestock products.[14] Together these commodity groups account for at least three-quarters of agricultural gross domestic product (AgGDP) in most ECA countries. The first three columns of Table 5.1 report the growth rates of these subsectors over the past five to eight years. Assuming such growth rates (together with recent growth rates for other, smaller, agricultural sectors and two nonagricultural subsectors) and using the multimarket model to project these rates to 2015, the annual growth rates for AgGDP and overall GDP are obtained (columns 4 and 5 of Table 5.1).

These business-as-usual outcomes suggest that in all countries except Sudan and Uganda (assumed to continue to register relatively high growth rates as they recover from civil strife), AgGDP and overall GDP will grow at rates below the 3 percent required to keep pace with population growth. Per capita GDP growth rates would therefore be less than 1 percent in a majority of countries. Kenya's per capita GDP growth to 2015 would be essentially zero; those of Madagascar, Rwanda, and Tanzania would be only marginally higher. Burundi, Democratic Republic of Congo (DRC), Eritrea, and Ethiopia would register negative growth rates.

Clearly, with business-as-usual in agriculture, ECA's future would not feature broad-based economic growth. The United Nations Millennium Development Goals (MDGs) would be unattainable, as would other development goals identified by ECA countries—such as increased food and nutrition security. Indeed, with business-as-usual, the gap between demand and supply of major food crops in ECA would widen. For cereals, the supply shortfall would increase to 6 million metric tons by 2015, 50 percent more than that in 2003 and 15 percent of the total regional demand.

---

[14]Staples include cereals and root crops; cash crops include traditional exports (such as coffee and tea) and nontraditional exports (such as fruits and vegetables); livestock products include meats, milk, eggs, and skins and hides.

Table 5.1 Sectoral growth rates to 2015 in the business-as-usual scenario (%)

	Staples	Cash crops	Livestock	Agricultural GDP	GDP per capita
Burundi	2.43	2.26	0.18	1.77	–0.11
DRC	0.70	1.05	1.73	1.54	–2.29
Eritrea	1.26	0.77	0.76	1.18	–1.30
Ethiopia	1.63	2.60	4.79	2.19	–0.16
Kenya	2.14	1.15	4.91	2.39	0.04
Madagascar	2.91	2.51	1.06	2.93	0.25
Rwanda	3.92	3.12	4.28	3.63	0.27
Sudan	5.33	3.06	1.99	3.33	1.19
Tanzania	2.94	3.39	3.45	2.97	0.78
Uganda	3.56	2.24	5.06	4.19	1.35

The business-as-usual outcome sheds light on the largely disappointing results in ECA of agricultural development policies in the 1980s and 1990s that concentrated primarily on reducing impediments to trade in agricultural markets. Consider the implications for AgGDP and overall GDP growth of 50 percent reductions in domestic marketing margins and barriers to international trade (Table 5.2). Trade liberalization mainly affects farmers who produce internationally traded commodities. In general, the aggregate effect of trade liberalization on AgGDP and overall GDP is smaller than that from reductions in domestic marketing costs. Reductions in domestic marketing costs benefit all farmers—not only those who produce for domestic markets but also those who produce for international markets.[15] Note the limited effects of both trade liberalization and reductions in domestic marketing costs on GDP and AgGDP growth rates, neither of which would differ significantly from those in the business-as-usual scenario. In the absence of productivity growth in the sector, the agricultural supply response to reduced impediments to trade would be weak. The business-as-usual scenario thus captures the outcome of continued "blind" liberalization of agricultural markets in ECA. Reducing distortions and impediments to trade in agriculture was certainly necessary, but it was far from sufficient. More strongly, those ECA governments and donor agencies that surmised that "letting agricultural markets work" meant assigning peripheral roles to public sectors in agricultural development were, at the very least, misguided. As is shown below, achieving the sustained levels of productivity growth required to significantly raise incomes and reduce poverty implies important roles for public sectors, not least in helping markets work in ways that benefit the poor.

The remaining sections of this chapter explore the nature of "business not as usual" in ECA agriculture, and the implied role of the public sector in shaping a more positive future for ECA agriculture than that associated with business-as-usual.

Growth Options in Agricultural Subsectors

The New Partnership for Africa's Development estimates that to achieve the MDG of halving poverty by 2015, African countries must register overall economic growth rates in excess of 6 percent per year over the next 12 years. In economies dominated by

[15]Reducing marketing margins typically requires significant investment in transportation, including rural roads and recurrent road maintenance, as these costs represent the bulk of total marketing costs (Kherallah et al. 2002). Policy gains, although not easy, require far lower costs. Clearly both interventions benefit farmers and consumers.

Table 5.2 Effects on growth of 50 percent reductions in domestic marketing costs and trade barriers (%)

	GDP			Agricultural GDP		
	Base	Domestic	Trade	Base	Domestic	Trade
Burundi	2.1	2.8	2.2	1.8	2.6	1.9
DRC	0.5	0.5	1.4	1.5	1.0	0.3
Eritrea	1.4	2.1	1.6	1.2	1.8	1.0
Ethiopia	2.5	3.3	2.6	2.2	2.7	2.0
Kenya	2.5	3.0	2.6	2.4	3.1	2.6
Madagascar	3.4	4.0	3.4	2.9	3.4	2.5
Rwanda	3.7	4.6	4.0	3.6	3.2	2.5
Sudan	3.6	4.2	3.7	3.3	3.6	3.3
Tanzania	3.6	4.4	3.9	3.0	3.5	3.0
Uganda	4.1	4.8	4.3	4.2	4.1	3.5

Note: Base is the business-as-usual scenario.

agriculture, such as those in ECA, achieving such GDP growth rates means generating rapid growth in agriculture. But the various agricultural subsectors probably make different contributions to overall economic growth. And for given overall growth rates, growth in different agricultural subsectors can variously affect poverty reduction. Achieving rapid growth and poverty reduction in ECA therefore requires an understanding of which agricultural subsectors can most effectively drive growth and slash poverty in the region.

Subsectoral Priorities

On which agricultural subsectors might ECA governments depend for enhanced growth and poverty reduction, and why? To answer these questions, the multimarket model is used to explore the cumulative impact on AgGDP and GDP of equal percentage increases in each subsector's annual growth rate to 2015. Figures 5.1 and 5.2 summarize the results for 15 commodities (or commodity groups).[16]

Milk emerges as the most important commodity subsector for both AgGDP and GDP generation. Oilseeds, cassava, and fruits and vegetables also rank highly in both regards. The cases of maize and bananas, on one hand, versus oilseeds, sorghum and millet, and beef, on the other, are highly instructive of the underlying dynamics.

Cumulative AgGDP gains from growth in the maize and banana subsectors are larger than are those from growth in oilseeds, beef, and sorghum and millet. The reverse is true for GDP gains.

Consider maize versus sorghum and millet. With sustained productivity growth, maize becomes an exportable commodity. Its price within the region is therefore derived directly from world prices and thus unresponsive to changes in regional demand. Conversely, even with sustained productivity growth for sorghum and millet, these commodities remain nontradable. One might assume that growth in an exportable crop subsector that does not face a market constraint—such as maize here—should generate greater economywide gain (that is, gain in total GDP) than a similar level growth in a nontraded crop subsector—such

[16]The results summarized in Figures 5.1 and 5.2 assume 1 percent increases in growth rates. Higher (or lower) increases would have yielded identical relationships to those shown in the charts. The absolute levels of impact are less important than are the relative levels across commodities. Results are shown for only 15 commodities or commodity groups because those for the others are insignificant by comparison.

Figure 5.1 Cumulative agricultural GDP gains to 2015 from 1 percent additional growth in selected commodity subsectors (US$ millions)

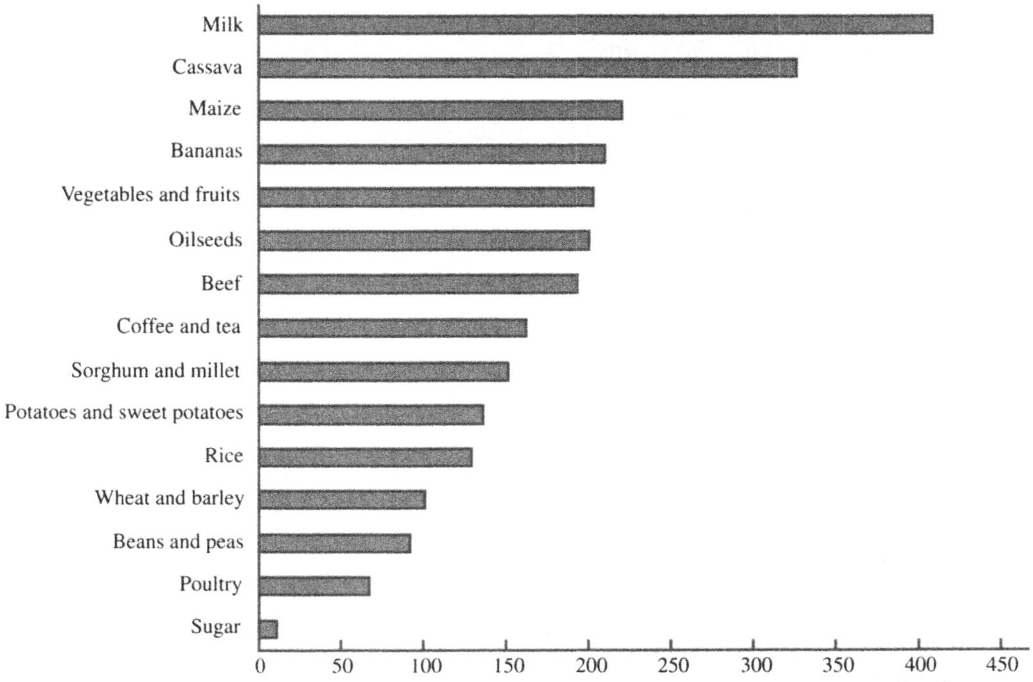

Figure 5.2 Cumulative GDP gains to 2015 from 1 percent additional growth in selected commodity group subsectors (US$ millions)

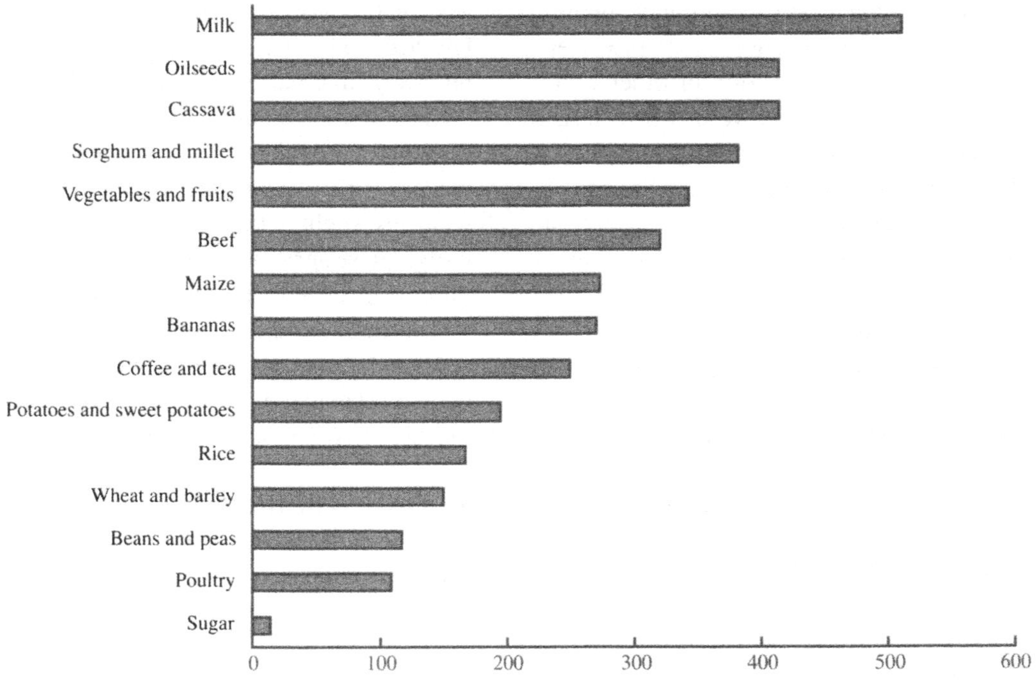

Source: Authors' simulations with multimarket model.

Figure 5.3 Cumulative GDP gains to 2015 from 1 percent additional growth in selected commodity group subsectors (US$ millions)

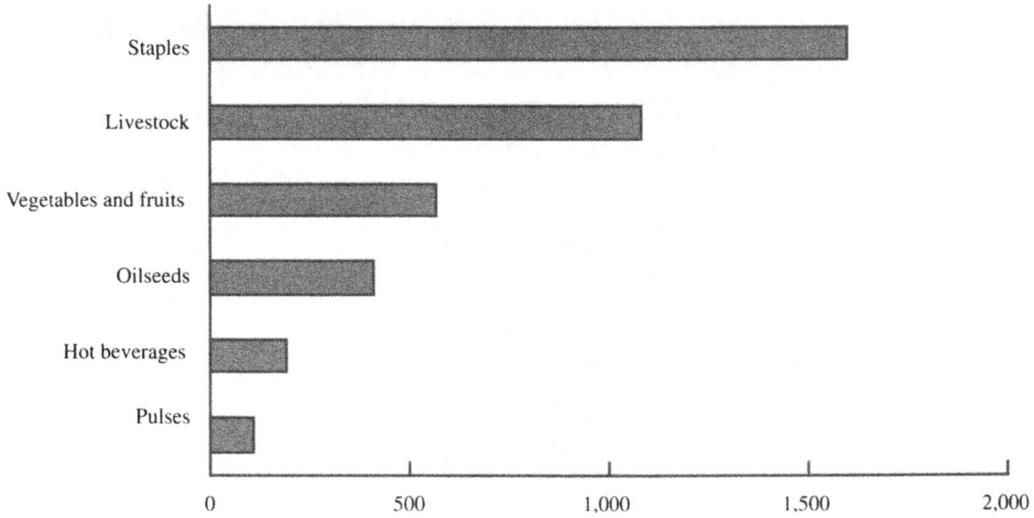

Source: Authors' simulations with multimarket model.

as sorghum here. However, that may not necessarily be the case if linkage effects are taken into account. For a nontraded crop, such as sorghum, higher supplies cause their prices to fall. This drop leads not only to higher demand for sorghum and millet, but also—crucially—to opportunities for reallocation of consumers' income to other items produced in the region. Reallocation, in turn, leads to supply responses from producers engaged in the production of other crops or commodities, which results in a larger economywide effect. By taking into account both direct effects and indirect linkage effects, the calculations show that growth in the sorghum subsector in fact generates higher gains in overall GDP than does a similar level of growth in the maize subsector.[17]

Similar dynamics underpin the relatively higher effects on GDP of growth in the oilseed and beef subsectors. Demand for these items tends to increase faster than growth in incomes. Such growth in demand permits sustained productivity growth without negative price effects, and thus, as described above for sorghum and millet, contributes to higher overall real incomes.

Taken together the staples subsectors result in the largest GDP gains, followed by livestock products, vegetables and fruits, and oilseeds (Figure 5.3). Higher growth for such hot beverages as coffee and tea—traditional export subsectors—results in relatively low GDP gains. Such crops as wheat, barley, rice, and pulses (beans and peas) have relatively high current and expected future demand in the region (Table 5.3). However, their production bases are relatively small, rendering their aggregate effects on GDP relatively slight. These results suggest that the greatest agriculture-led growth opportunities in ECA reside in commodities for which, first, there is a relatively large production base to begin with, and, second, there is a large and growing demand within the region. We return to this point below.

MDG Growth Targets

What would it take for ECA countries to achieve overall economic growth rates re-

[17]The consumption linkages generated by increased rural income are the most important linkages in low-income countries (see Vogel 1994).

Table 5.3 Supply, demand, and net supply of selected agricultural commodities, 2003, 2009, and 2015 (000s metric tons)

	2003			2009			2015		
Commodity	Supply	Demand	Net supply	Supply	Demand	Net supply	Supply	Demand	Net supply
Maize	10,546	10,803	–257	12,508	12,709	–201	14,968	15,032	-64
Rice	2,558	3,069	–511	2,954	3,691	–737	3,424	4,470	–1,046
Wheat	2,015	5,026	–3,011	2,285	5,846	–3,561	2,602	6,846	–4,244
Sorghum	5,270	5,321	–51	6,204	6,336	–132	7,292	7,572	–280
Barley	1,060	1,092	–32	1,137	1,264	–127	1,217	1,467	–250
Millet	1,701	1,702	–1	2,003	2,007	–4	2,367	2,374	–7
Oats	55	55	0	65	65	0	77	77	0
Other cereal	1,690	1,681	9	1,859	1,862	–3	2,053	2,057	–4
Potatoes	3,137	3,181	–44	3,777	3,885	–108	4,582	4,763	–181
Sweet potatoes	6,426	6,426	0	7,916	7,915	1	9,798	9,791	7
Cassava	30,387	30,386	1	34,199	34,235	–36	38,814	38,886	–72
Other roots	4,483	4,481	2	5,029	5,026	3	5,654	5,649	5
Beans	1,359	1,330	29	1,471	1,463	8	1,626	1,725	–99
Peas	1,339	1,416	–77	1,552	1,656	–104	1,817	1,939	–122
Groundnuts	1,474	1,459	15	1,790	1,703	87	2,205	2,019	186
Sesame seed	336	197	139	396	241	155	470	298	172
Other oil crops	2,052	2,034	18	2,482	2,428	54	3,447	3,263	184
Vegetable oil	330	961	–631	355	1,120	–765	391	1,329	–938
Raw sugar	1,581	1,887	–306	1,874	2,235	–361	2,249	2,682	–433
Vegetables	9,844	8,666	1,178	10,915	10,148	767	12,181	11,952	229
Bananas	15,335	15,334	1	17,825	17,824	1	20,825	20,825	0
Fruits	3,940	3,789	151	4,568	4,303	265	5,334	4,904	430
Coffee	681	208	473	770	243	527	875	285	590
Tea	351	52	299	370	61	309	394	73	321
Spices and beverages	6,326	6,326	0	7,376	7,376	0	8,580	8,726	–146
Cotton lint	168	86	82	201	102	99	243	122	121
Beef	1,290	1,294	–4	1,538	1,537	1	1,842	1,842	0
Mutton	432	425	7	463	474	–11	500	526	–26
Poultry eggs	468	484	–16	567	595	–28	691	744	–53
Other meat	573	574	–1	611	659	–48	656	764	–108
Milk	9,089	9,266	–177	11,437	11,928	–491	14,552	15,588	–1,036
Fish	1,232	1,230	2	1,435	1,374	61	1,688	1,559	129

quired to attain the MDG for poverty reduction? What levels of growth would be required of specific agricultural subsectors? To answer these questions, consider four agricultural commodity groups: staples (cereals, roots, and tubers), livestock products, nontraditional exports (fruits and vegetables), and coffee. The first two subsectors dominate agriculture in most countries. The last two are key export subsectors, which, though small compared to the first two, either have potentially large scope for growth (nontraditional exports) or loom large in total agricultural export earnings in many countries (coffee).[18] Suppose each of these agricultural subsectors were to grow at a rate that in itself (that is, with business-as-usual assumptions for each of the other three sectors) yielded an overall growth rate for the economy that filled the gap between the business-as-usual growth rate and the 6 percent required to achieve MDG for poverty

[18]The exercise described here was undertaken for other commodity groups. The outcome reported here also applies to those groups.

Table 5.4 Sectoral growth rates with improved investment strategies (%)

	Staples	Cash crops	Livestock	Agricultural GDP	GDP	GDP per capita
Burundi	6.8	7.1	5.2	4.3	5.6	3.4
DRC	6.3	7.8	8.9	4.3	4.6	1.4
Eritrea	6.9	6.3	6.8	5.9	6.4	3.6
Ethiopia	5.2	6.7	9.1	5.4	6.0	3.4
Kenya	5.9	5.3	9.2	6.6	6.3	3.8
Madagascar	5.7	5.4	4.1	5.6	6.3	3.1
Rwanda	6.7	6.1	7.4	5.4	6.6	3.0
Sudan	8.1	6.1	5.1	5.6	6.6	3.3
Tanzania	5.8	6.5	6.6	5.7	6.6	3.8
Uganda	5.5	4.3	7.2	6.0	6.6	3.6

reduction. Suppose further that these subsectoral growth rates were combined with similarly targeted nonagricultural growth rates. The scenario in Table 5.4 would result. Most ECA countries would achieve GDP growth rates of 6 percent or more to 2015, and per capita GDP growth rates in excess of 3 percent. The only exception would be DRC, which would take longer to recover from its negative starting growth rate in 2003. Achieving GDP growth rates required to meet MDG poverty reduction targets implies threefold increases in agricultural sectoral and subsectoral growth rates.

Food Supply and Demand

Not surprisingly, growth such as that reported in Table 5.4 would significantly improve the region's food supply situation. In contrast to the business-as-usual outcome in which food imports would rise by more than 50 percent by 2015, with the expanded growth in staples subsectors, the region would generate a food surplus by 2015 equivalent to 3.5 percent of total regional consumption by 2015 (Figure 5.4). Most of this surplus would come from increased maize supplies, which would progressively outstrip maize demand because of the relatively low income-responsiveness of that demand. Once again, this result highlights the importance of the demand side. Without growth in other sectors (especially in the livestock subsector to generate demand for maize as feed and in nonagricultural sectors to help raise incomes, increase consumption of livestock products, and further spur demand for maize and other staples as feed) growth in the staples subsectors would be constrained. More positively, if growth in the staples subsectors were combined with that in the livestock subsector, other agricultural subsectors, and in nonagriculture, the potentially negative price effects of growth in the staples subsectors could be avoided while enhancing overall food availability in the region.

ECA is highly dependent on wheat imports. In the business-as-usual scenario, wheat imports would rise by 40 percent by 2015, accounting for more than 60 percent of demand. Unlike maize, however, demand-side constraints would not limit growth in this subsector. Expanded productivity in the wheat subsector would reduce wheat imports without depressing prices. But given the relatively small supply base for wheat in the region (see Table 5.3), the higher production would not fully offset higher demand caused by rising incomes. The region would therefore remain a wheat importer, even with substantial growth in the wheat subsector. An analogous picture emerges for rice, where, with few demand-side constraints on productivity growth, the region could approach self-sufficiency by 2015.

Figure 5.4 ECA net imports of major cereals in 2003 and in alternative growth scenarios

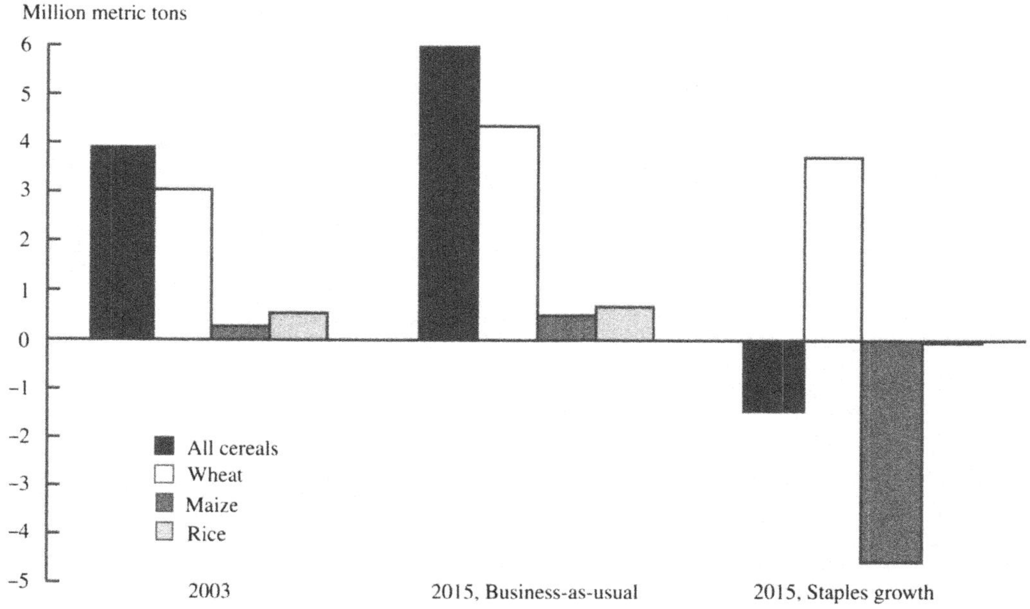

Source: Authors' simulations with the multimarket model.
Note: Negative values are net exports.

Growth Options for Poverty Reduction: Insights from Ethiopia

In assessing the implications of alternative growth options for poverty reduction, a key recognition is that effects vary according to the roles played by given commodities in household production, consumption, and trade, and thus also according to the potential for broad-based income growth through investments that target growth in particular agricultural subsectors. Understanding such potential requires detailed country-level analysis within a multimarket framework. Such an analysis was undertaken for Ethiopia, where requisite household-level income and poverty data are available. Ethiopia's agricultural characteristics are very similar to those of several countries in the region (see Tables 3.1, 3.3, 3.4, and 3.5). Insights emerging from the Ethiopian case may therefore be viewed as broadly representative for the region. The analytical horizon is once again set at 2015.

Growth

As in several other ECA countries, staples loom large in Ethiopia's agricultural sector. Growth in the staples subsectors would therefore spur faster overall growth to 2015 than would that in any other subsector. For instance, 1.5 percent of additional growth for staples would be equivalent to 9.0 percent of additional growth for nontraditional exportables, which constitute a dynamic but small agricultural subsector in Ethiopia. Specifically, a 1.5 percent increase in the growth rate of the staples subsectors would result in a 3.7 percent increase in the AgGDP growth rate, versus a 3.5 percent increase from a 9.0 percent increase in the growth rate for nontraditional exportables. Similar differences in GDP growth rates from such subsectoral expansions would also result.

Poverty Reduction

Growth in different subsectors has differential effects not only on overall growth but on poverty. For Ethiopia, assuming subsectoral

Figure 5.5 Differential poverty reduction in Ethiopia from growth in key agricultural subsectors

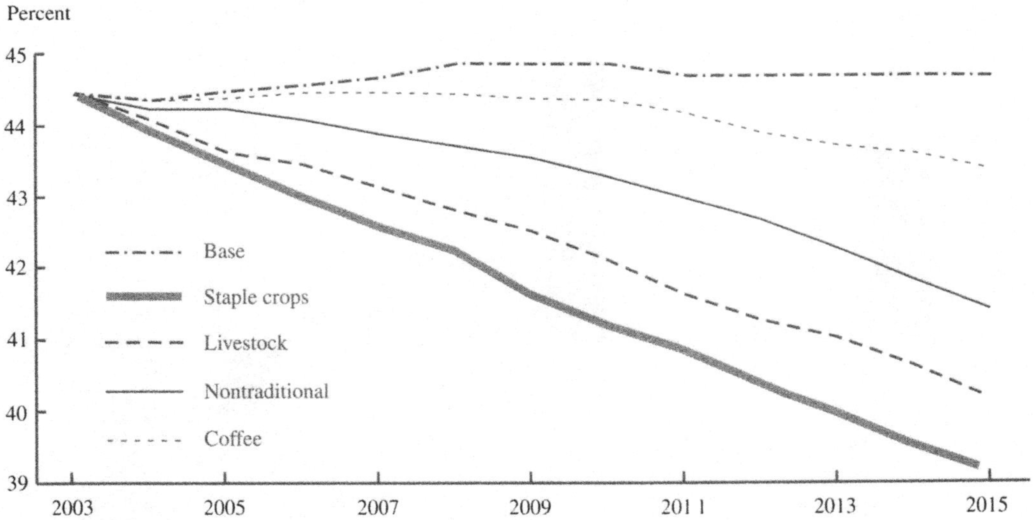

Source: Authors' simulations with the multimarket model.

growth rates for staples, livestock products, nontraditional exports, and coffee that induce an equivalent 4 percent growth rate in AgGDP yields markedly different effects on poverty (Figure 5.5). The largest reductions in poverty would come from growth in staples, with successively less change in the livestock, nontraditional exports, and coffee subsectors.

The significance of this result cannot be overstated. Growth in export subsectors—especially that in nontraditional export subsectors—is often put forward as a pathway out of poverty for such countries as Ethiopia and others in ECA. The current analysis indicates that such advice is highly misplaced.

Improving productivity in staple crop production would directly benefit the great majority of small farmers by easing key resource constraints in the activity to which they devote most of their resources, namely, staple food production. Demand-side considerations are also important. Staple crops often account for large shares of household expenditures and are therefore key sources of food energy for both rural and urban poor consumers. Household survey data from Ethiopia reveal that households whose incomes lie below the rural poverty line spend about 70 percent of their total income on staples, 30 percent higher than the rural average. In urban areas, such impoverished households spend almost 50 percent of their income on staples, 65 percent higher than the urban average. Growth in staples subsectors would therefore ameliorate both rural and urban poverty.[19]

Conversely, nontraditional exportables subsectors typically cover small groups of relatively wealthy and geographically concentrated farmers. This limited demographic and geographic scope for yielding broad-based income expansion is accentuated by key supply-side and demand-side constraints.

[19]Recall that staples include cereals and root crops; cash crops include traditional exports (such as coffee and tea) and nontraditional exports (such as fruits and vegetables); livestock products include meat, milk, eggs, and skins and hides. The bundle of staple crops thus varies across countries. In some countries (for example, Kenya), maize is key whereas in others (such as DRC), cassava is more important; some of these staples are more actively traded than are others. The key point is that these are crops that are widely produced and consumed by smallholders.

Figure 5.6 Differential poverty reduction in Ethiopia from agricultural and nonagricultural growth and improvements in agricultural markets

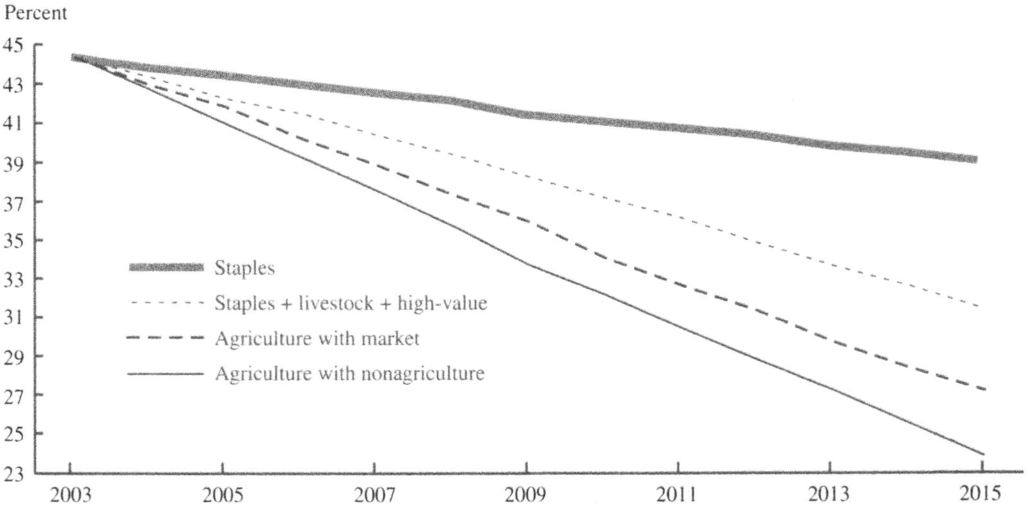

On the supply-side, the initial investments needed to meet stringent technical and financial requirements in export-oriented production and trade typically render such activities beyond reach for most smallholders. On the demand-side, increased production of most exports may provide little nutritional benefit to poor consumers in both rural and urban areas because such products are often intended for export markets.[20]

Nonagricultural Sectors and Markets

As noted above, on its own, agricultural growth is not sufficient to meet MDG poverty reduction targets. Growth in nonagricultural sectors and improvements in market conditions are also required. Growth in nonagricultural sectors not only provides off-farm income-earning opportunities for rural populations, it also generates market demand for agricultural products, further stimulating growth. Without improvements in market conditions, an increase in the supply of agricultural products may depress prices and reduce farm income. Figure 5.6 illustrates this argument. Combined with productivity growth in staples, livestock, and nontraditional exports, growth in nonagricultural sectors and improvements in markets would lead to reductions in poverty rates in Ethiopia equivalent to those in MDGs.

Subnational Differentiation

In most countries, poverty rates vary geographically. Growth options for poverty reduction should therefore take such variations into account. Consider Ethiopia's food deficit and food surplus areas. Productivity growth in staples subsectors would benefit the poor in food surplus areas the most (Figure 5.7). The poverty rate in food surplus areas would fall from 30 percent in 2003 to 23 percent in 2015 (a fall of 23 percent), whereas that in food deficit areas would fall from 62 percent in 2003 to 54 percent by 2015 (a fall of only 13 percent). In contrast, productivity growth in the livestock subsector would reduce poverty more sharply in food deficit areas than it would in food

[20]Note, however, that this is not to suggest that production of traditional export crops is unimportant as a source of income and employment in Ethiopia or elsewhere in ECA. Rather, the argument here relates to potential for *broad-based* impacts on growth and poverty reduction.

Figure 5.7 Subnational differences in poverty reduction in Ethiopia from growth in staples and livestock subsectors

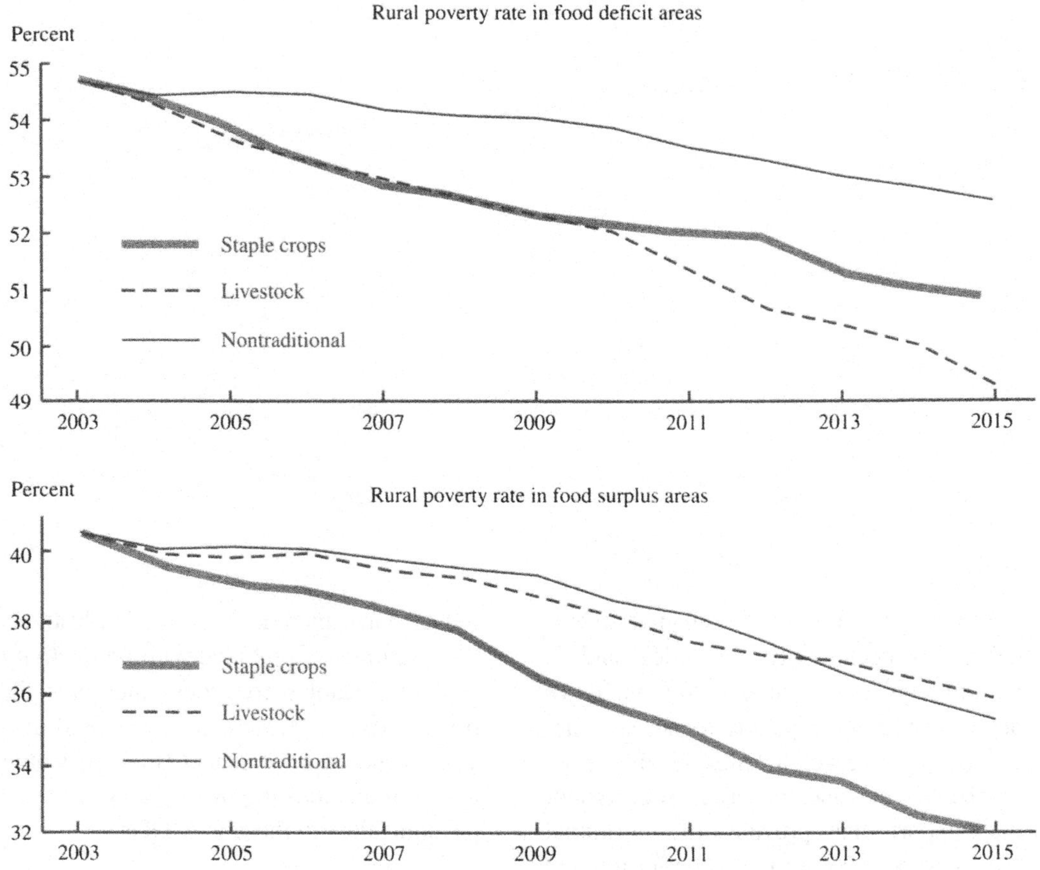

Source: Authors' simulations with the multimarket model.

surplus areas—from 62 percent to 53 percent in deficit areas (a 15 percent decline), and from 30.0 percent to 27.6 percent in surplus areas (an 8.0 percent fall).

Summary

This chapter outlines the ingredients of growth-enhancing, poverty-reducing agricultural development policy in ECA. The aim is to identify strategic priorities for agricultural development in the region that can assist national and regional stakeholders in defining their own priorities, objectives, strategies, and action plans.

Based on a methodology that integrates spatial analysis with economic modeling in a multimarket model of ECA agriculture, continuation of these trends—business-as-usual—has been shown to imply agricultural and overall growth rates inadequate to reduce poverty in the region. In the business-as-usual scenario, no ECA country would achieve the growth rates required to meet MDG of halving poverty by 2015. On the contrary, the growth rates would imply deepening poverty in the region.

The business-as-usual outcome sheds light on the largely disappointing results in ECA of agricultural development policies in the 1980s and 1990s that concentrated primarily on reducing impediments to trade in agricultural markets (Kherallah et al.

2002).[21] Specifically, in the absence of agricultural productivity growth, both trade liberalization and reductions in domestic marketing costs result in GDP and AgGDP growth rates that differ only slightly from those in the business-as-usual scenario. The assumption that "letting agricultural markets work" meant assigning peripheral roles to public sectors in agricultural development was incorrect.

Further analysis yielded numerous insights into the nature of agricultural development that might allow countries to avoid business-as-usual outcomes. These findings are summarized here and in Table 5.5 for the case of Ethiopia, the ECA country for which the most complete data are available:

- Achieving GDP growth rates required to meet MDG poverty reduction targets implies threefold increases in agricultural sectoral and subsectoral growth rates.
- Although growth in export subsectors is often put forward as a pathway out of poverty for countries in ECA, the greatest reductions in poverty would come from growth in subsectors for which demand is highest within the region—staples, livestock products, oilseeds, and fruits and vegetables.
- Balanced growth strategies featuring growth in many agricultural subsectors lead to higher overall economic growth than does that featuring growth in a few sectors.
- Agricultural productivity growth alone is insufficient to meet MDG poverty reduction targets. Growth in nonagricultural sectors and improvements in market conditions are required.
- Because poverty rates vary geographically within countries; growth strategies that take such differences into account lead to larger reductions in poverty than do those that ignore such differences.

[21]The Kenyan case is illustrative. Before market liberalization policies were instituted in the early 1990s, statutory marketing bodies played unusually dominant roles in domestic agricultural markets. High administrative costs and delayed payments to farmers—combined with domestic movement controls for farm outputs, import restrictions on farm inputs, and price controls on both outputs and inputs—imposed heavy burdens on farmers and traders. Market liberalization lifted these controls but with only limited effects on productivity. Some authors argue that Kenya (and many other countries in Africa) never actually moved to a liberalized market environment, and that many of the most fundamental elements of the reform programs remain unimplemented, were reversed within several years, or were implemented in such a way as to negate private sector investment incentives (Jayne et al. 2002).

Table 5.5 Summary of alternative growth, nutrition, and poverty scenarios in ECA: Ethiopia

	Business-as-usual	Growth in staples only	Growth in livestock only	Growth in nontraditionals only	Growth in coffee only	Growth in staples, livestock, and nontraditionals	Reductions in marketing costs	Halving poverty by 2015
Agricultural GDP growth rate (%)	2.8	3.8	3.9	3.6	3.5	5.5	5.7	5.7
GDP growth rate (%)	3.2	4.0	4.1	3.7	3.6	5.3	5.9	6.5
Calories per capita by 2015 (base = 1,834)	1,744	1,999	1,840	1,808	1,748	2,176	2,215	2,288
Malnourished children by 2015 (base = 47%)	48.9	43.7	46.9	47.5	48.8	40.5	39.9	38.6
Poverty rate by 2015 (base = 44.5%)	46.2	37.7	40.0	41.0	42.8	28.4	24.2	19.8

CHAPTER 6

Strategic Priorities for Agricultural Development

Setting priorities requires comparing and ultimately choosing among alternative courses of action. At issue here are the potential (or foregone) contributions of given choices to growth and poverty reduction in eastern and central Africa (ECA). Using contributions to gross domestic product (GDP) and agricultural GDP (AgGDP) to gauge agriculture-based contributions to growth and poverty reduction, this chapter identifies priorities in two dimensions: (1) among major commodity subsectors at regional and national levels and (2) across agricultural development domains. The picture that emerges is mixed. Although priorities across agricultural development domains are clear, those among commodities at regional and national levels are less so. The potential benefits for regional coordination of agricultural research and development (R&D) are quantified and found to be substantial.

Priority Commodities for Countries and Country Groups

As detailed in Chapter 5 (see Figures 5.1, 5.2, and 5.3), when ECA is viewed as region, milk emerges as the most important commodity subsector for growth in both AgGDP and GDP. Oilseeds, cassava, and fruits and vegetables also rank highly. Given the diversity of biophysical and socioeconomic conditions across ECA countries, these regional priorities among commodities may not apply for individual countries. Figures 6.1 and 6.2 show the distributions of growth-induced GDP gains across countries for given commodities and across commodities for given countries, respectively. Countries capture different shares of GDP gains accruing to given commodity subsectors (for example, milk, cassava, rice, and wheat and barley in Figure 6.1), and some country gains are dominated by given commodities (for example, Democratic Republic of Congo [DRC], Madagascar, and Rwanda in Figure 6.2).

Table 6.1 shows the result of an exercise that computes the divergence of national distributions of commodity gains from the regional distribution. Kenya, Tanzania, and Uganda form a group whose priorities closely match those for the region as a whole. Regional priorities would appear to be less relevant to Eritrea, Ethiopia, and Sudan, where milk and cereals are considerably more important than elsewhere in the region. Regional priorities seem to be least relevant to Burundi, DRC, Madagascar, and Rwanda. In Burundi and Rwanda, bananas, potatoes, and sweet potatoes are crucial. In DRC, oilseeds and cassava are paramount. In Madagascar, rice is central.

Table 6.2 shows the results of a computation of divergences of gains accruing to countries from an equal distribution of gains across countries for each commodity. Although distinct clusters of commodities do not emerge, a clear gradation is evident, from commodities whose distributions across countries are relatively equal (such as fruits and vegetables) to those

Figure 6.1 Country shares of commodity GDP gains

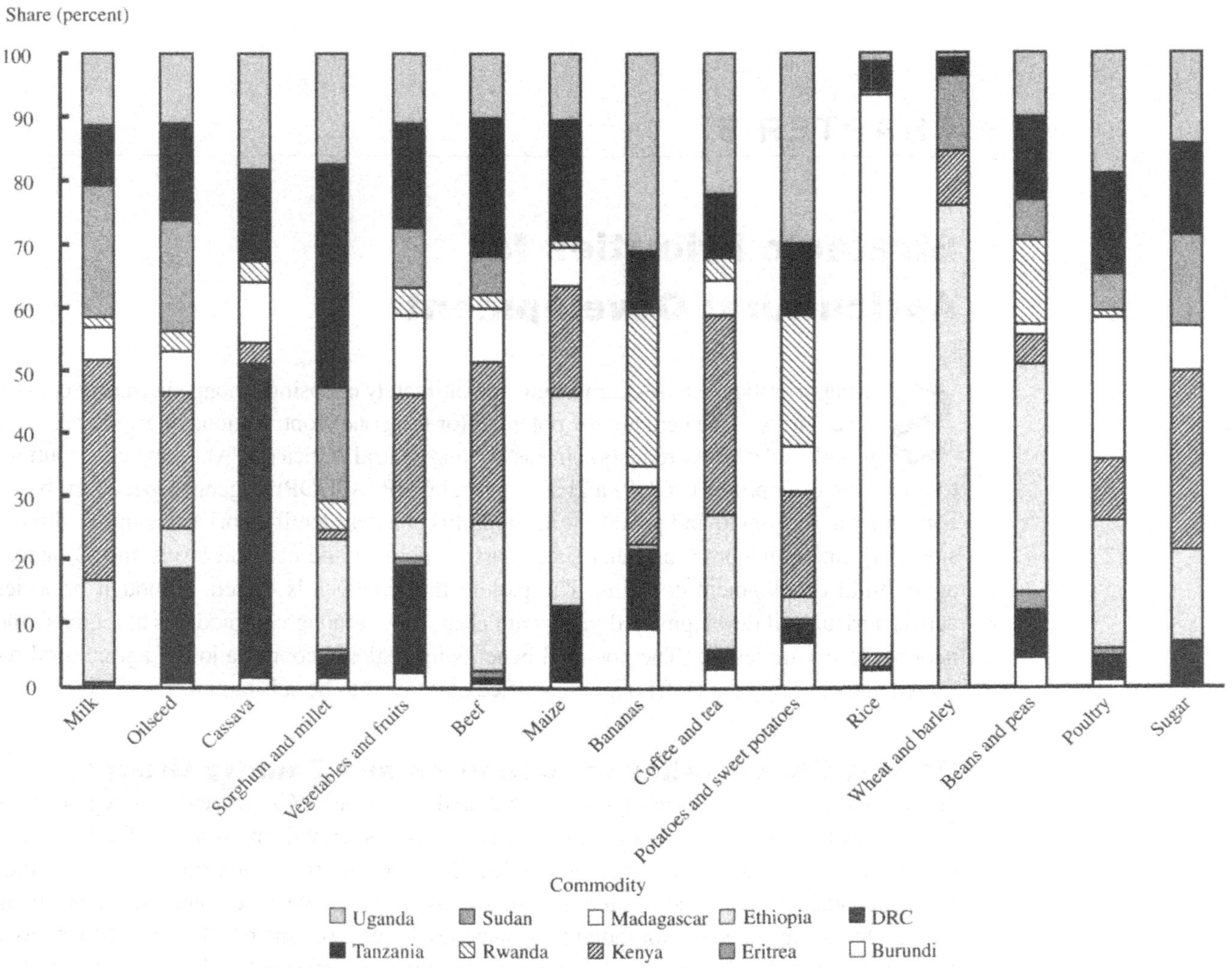

whose distribution is relatively unequal (such as cassava). The results for cassava, milk, and sorghum and millet are especially meaningful. Recall that these commodities ranked very highly for the region as a whole (see Figure 5.2). But the result in Table 6.2 suggests that such gains would be concentrated in a few countries. Figure 6.1 reveals that for cassava, gains would be concentrated in DRC, Tanzania, and Uganda; for milk they would fall largely in Ethiopia, Kenya, and Sudan; and for sorghum and millet they would accrue mainly to Ethiopia, Sudan, Tanzania, and Uganda. Fruits and vegetables, beef, oilseeds, and maize emerge as commodity subsectors in which growth would result in gains that are both large and widespread. This conclusion is confirmed in Table 6.3, which shows the results of an analysis in which commodity subsectors are ranked, first, on the size of the GDP gains they induce, and, second, on the distribution of those gains across countries in the region. In both rankings, commodity subsectors are ranked from 1 to 15, with 15 assigned to subsectors that yield the highest size and greatest spread of gains, and 1 to the lowest. Oilseeds, fruits and vegetables, beef, and maize are superior on both accounts. Sorghum and millet generate high gains on aggregate but exhibit only moderate regionality of those gains. Coffee, tea, pulses (beans and peas), and poultry generate relatively low gains on aggregate but have relatively high

Figure 6.2 Commodity shares of country GDP gains

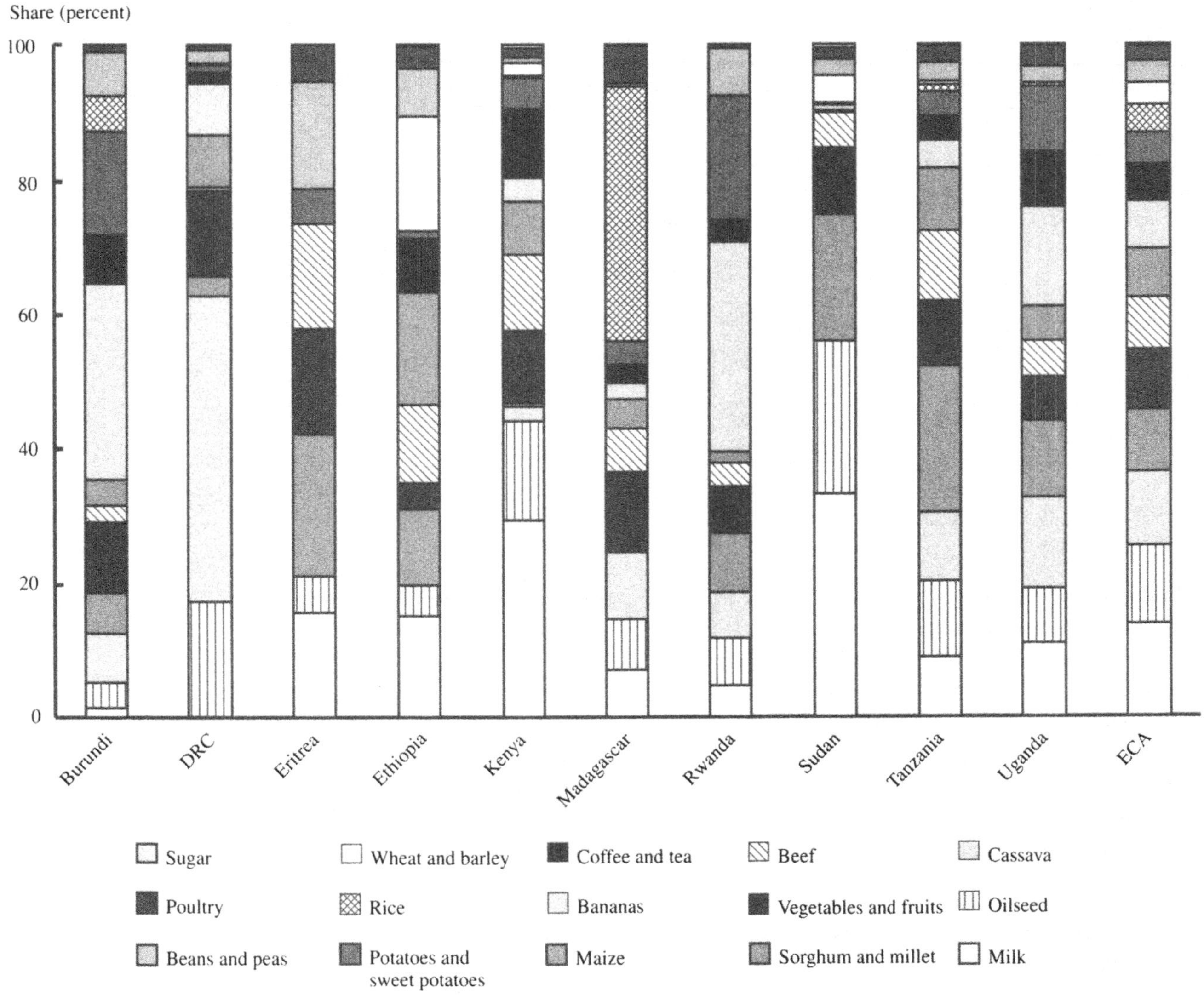

regionality scores and thus, based on this criterion, they may be accorded higher priority than would otherwise be the case. Milk and cassava generate high gains overall but have low regionality scores, suggesting sharp tradeoffs between growth and equity from regional development investments in these subsectors.

Priorities across Agricultural Development Domains

As before, annual growth rates in each of the commodity subsectors included in the multimarket model are increased by 1 percent and the multimodel is solved to 2015 to predict GDP. The resultant GDP gains summed over all 10 countries and distributed across the 8 agricultural development domains (according to shares of commodities produced in each domain; see Table 6.4) are summarized in Figure 6.3.

The high agricultural potential, low market access, low population density (HLL) domain emerges as having by far the highest growth potential in the region. As noted in Chapter 3, this domain accounts for close to 30 percent of regional population and almost 40 percent of land area, cropped area, pasture area, and livestock population. Large segments of all ECA countries fall in

Table 6.1 Divergences from the regional distribution of commodity GDP gains to 2015

Country	Divergence
Tanzania	0.03
Uganda	0.03
Kenya	0.05
Ethiopia	0.09
Sudan	0.10
Eritrea	0.11
Burundi	0.16
DRC	0.17
Rwanda	0.19
Madagascar	0.26

Note: The divergence is computed as the sum of squared differences between national shares of commodity GDP gains and the corresponding regional shares.

Table 6.2 Commodity subsector divergences from an equal distribution across countries of GDP gains to 2015

Commodity	Divergence
Vegetables and fruits	0.01
Beef	0.03
Oilseed	0.03
Maize	0.04
Coffee and tea	0.04
Beans and peas	0.05
Potatoes and sweet potatoes	0.05
Poultry	0.06
Sorghum and millet	0.06
Wheat and barley	0.08
Sugar	0.09
Milk	0.12
Bananas	0.13
Rice	0.15
Cassava	0.16

Note: The divergence is computed as the sum of squared differences between commodity shares of national GDP gains and an equal distribution across all countries in the region.

this domain, and large shares of all commodities are produced therein (Table 6.4). Complications raised by potential growth-equity tradeoffs are therefore likely to be minimal from growth originating in the HLL domain. This domain's low population density also suggests relatively low pressure on natural resources, and thus potentially few tradeoffs between growth and sustainability.

The remaining seven domains can be grouped into two sets: (1) the LLL, HHH, and HLH domains and (2) the LHH, HHL, LLH, and LHL domains (see Table 3.8 for

Table 6.3 Commodity rankings taking both size and distribution of gains into account

Commodity	Size score	Distribution score	Total
Oilseeds	14	13	27
Vegetables and fruits	11	15	26
Beef	10	14	24
Maize	9	12	21
Sorghum and millet	12	7	19
Milk	15	4	19
Coffee and tea	7	11	18
Potatoes and sweet potatoes	6	9	15
Cassava	13	1	14
Beans and peas	3	10	13
Bananas	8	3	11
Poultry	2	8	10
Wheat and barley	4	6	10
Rice	5	2	7
Sugar	1	5	6

Note: The size and distribution scores range from 1 to 15, with 15 assigned to subsectors that yield the highest size and greatest spread of gains, and 1 to the lowest.

Table 6.4 Production shares of selected agricultural commodities across agricultural development domains (%)

Domain	Milk	Cassava	Maize	Bananas	F&V	Oilseeds	Beef	C&T	S&M	P&SP	Rice	W&B	B&P	Poultry	Sugar
HHH	19	8	10	21	19	7	3	9	5	8	4	16	17	19	18
HHL	6	4	5	8	6	4	3	6	6	4	6	5	7	6	8
HLH	15	7	11	16	15	9	3	11	5	7	3	22	13	15	10
HLL	23	40	32	27	23	39	39	46	57	40	47	20	30	23	30
LHH	11	3	6	12	11	3	1	2	2	3	2	4	10	11	4
LHL	4	2	3	2	4	2	2	1	2	2	3	2	2	4	4
LLH	5	1	3	2	5	3	1	4	2	1	2	8	2	5	3
LLL	12	22	18	4	12	16	25	12	13	22	22	10	10	12	16

Notes: Domain refers to agricultural potential/market access/population density—for example, HLL = high potential, low access, low density. H = high; L = low; F&V = fruits and vegetables; C&T = coffee and tea; S&M = sorghum and millet; P&SP = potatoes and sweet potatoes; W&B = wheat and barley; B&P = beans and peas. The allocation of commodity production to development domains was completed as follows: cassava, maize, bananas, coffee, sorghum, millet, rice, wheat, barley, beans, peas, and sugar were based on district-level crop allocation data for Africa. Vegetables, fruits, poultry, and milk were allocated across domains using shares of population. Beef was allocated by shares of pasture area. Oilseeds and tea were allocated using shares of crop area.

Figure 6.3 Cumulative GDP gains to 2015 by agricultural development domain from 1 percent additional growth in commodity subsectors (US$ millions)

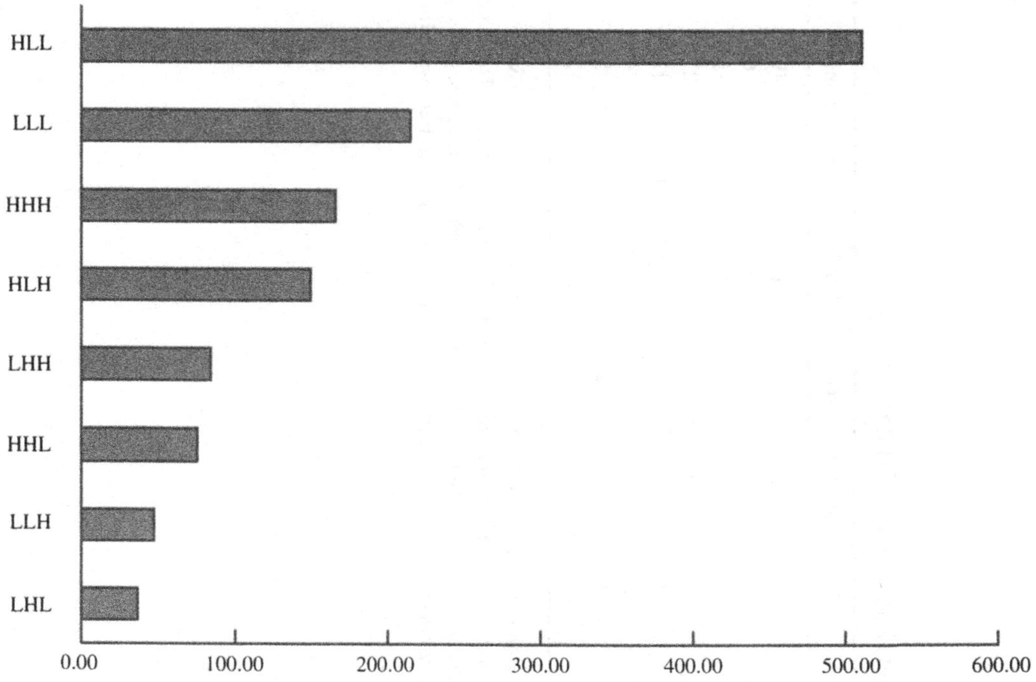

the definitions of these domains). The three domains in the first set are clearly of lower priority than the HLL domain, but they are also clearly of higher priority than the four domains in the second set. The LLL, HHH, and HLH domains can therefore be considered as moderately high priority; the LHH, HHL, LLH, and LHL domains are low priority.

Among the three moderately high priority domains, the relatively high GDP gains generated in the LLL domain reflect the prominence of livestock in this zone. Recall from Chapter 5 that milk and beef rank high among commodities in this respect, as do cassava and maize, both of which are important in this domain. The HHH and HLH zones produce relatively large gains because of their importance as suppliers of milk, poultry, bananas, fruits and vegetables, wheat, barley, and legumes.

Growth in the LLL, HHH, and HLH domains raises the possibility of tradeoffs between growth and sustainability. In LLL zones, concerns arise from the fragile and uncertain environments; in HHH and HLH areas, concerns spring from high population pressure. In both cases, productivity gains may be possible only at relatively high cost.

The relatively high potential and low population density in HLL areas implies a less sharp tradeoff between growth and sustainability in these zones than in the LLL, HHH, and HLH domains. In the low density but low potential LLL areas, livestock systems undergird livelihoods. Improved animal health, breeding for disease resistance and improved nutritive value, and improved pasture management are therefore crucial. Challenges in natural resource management are raised by the extensive nature of most livestock production systems in these areas and thus by the need for collective management of key resources. Development of property rights regimes that improve long-term incentives to invest in, sustain, and improve these resources is therefore important. Where irrigation is an option and cropping feasible, similar institutional challenges are raised by the likely need for community management of irrigation schemes. In such cases, new institutional arrangements will probably be required to develop new rules governing claims to water resources and to determine the means by which water resources are managed by individuals or groups.

In HHH and HLH zones, natural ecosystems are under pressure from persistently low levels of agricultural productivity growth coupled with growing food demand. Productivity growth will hinge on solutions to such problems as soil nutrient depletion, soil erosion, pests, and weeds. Many of the technologies required to address these problems already exist. Most of these technologies are knowledge intensive, implying the need for structures and processes that promote sustained learning among not only farmers but also service providers likely to be involved in successful technology adoption. The difficulties raised by soil nutrient depletion in these areas cannot be overstated. Again, sustainable solutions will likely hinge on institutional innovations—in this case those that lead to efficient and sustainable demand and supply of fertilizers and improved seeds, along with shifts in production practices toward greater commercialization.

The four low priority domains—LHH, HHL, LLH, and LHL—are progressively smaller in size and agricultural importance (Tables 3.4 and 4.1) and thus result in progressively smaller GDP gains from growth in commodity subsectors. Agriculture-based growth in these domains is unlikely to be large enough to warrant major agricultural development investments. Best-bet growth-enhancing options in these areas are likely to reside outside agriculture.

To build further understanding of priorities across the eight agricultural development domains, suppose that the domains are ranked from 1 to 8, depending on their effects on GDP gains to 2015 (as shown in Figure 6.3), with the domain producing the largest impact assigned 8, the lowest 1. Suppose also that the 15 commodities in Figure 5.2 are similarly ranked from 1 to 15, depending on their effects on GDP gains, with the commodity giving the largest GDP gain assigned 15, the lowest 1. And suppose that the share of a commodity produced in a given domain (for example, 19 percent for milk in the HHH domain, or 40 percent for cassava in the HLL domain; Table 6.4) is scaled by the product of the two above-mentioned rankings. The result is an index that captures the importance of a commodity or domain in ECA conditioned by its potential for generating growth in the region. Thus for milk in the HHH domain, the index is $15 \times 6 \times 0.19 = 17.12$. For cassava in the HLL domain, the index is $13 \times 8 \times 0.40 = 41.6$.

The resulting priority ranking of domains is unchanged from that reported in Figure 6.3 (Table 6.5). For commodities, an important change is the emergence of sorghum and millet in third spot, replacing cassava, which falls to fourth (Table 6.6). This result is driven by the large shares of sorghum and millet produced in the top two domains, HLL and LLL (57 and 13 percent, respectively; Table 6.4).

Table 6.7 shows the ranking for all commodity-domain combinations. As expected, the HLL domain dominates the high positions. However, note the emergence of sorghum and millet as the top priority commodities in this zone. And note that milk ranks only fifth in the HLL domain. But note also that among the top 20 commodity-domain combinations, milk appears four times. This frequency explains milk's overall priority as the highest regionwide. Finally, note the relatively high position of cassava in the LLL domain. Combined with the country-level priorities for commodities described in Chapter 4, these commodity-domain rankings provide a basis for more focused priority setting across commodities and domains than when the region is viewed as a whole. For instance, research on maize in the HLL zone (index = 23.04, rank = 7) may be easier to justify than that in the HHH domain (index = 5.51, rank = 34). Indeed, within the HHH domain, research on

Table 6.5 Domain rankings adjusted by commodity distributions

Domain	Score
HLL	342.72
LLL	129.61
HHH	80.29
HLH	60.30
LHH	26.73
HHL	18.34
LLH	7.55
LLL	2.98

Notes: Domain scores sum over commodities. Domain refers to agricultural potential/market access/population density—for example, HLL = high potential, low access, low density. H = high; L = low.

Table 6.6 Commodity rankings adjusted by domain distributions

Commodity	Score
Milk	80.34
Oilseeds	75.33
Sorghum and millet	75.29
Cassava	75.20
Fruits and vegetables	58.92
Beef	53.75
Maize	48.67
Bananas	42.56
Coffee and tea	41.56
Potatoes and sweet potatoes	34.71
Rice	30.13
Wheat and barley	19.59
Beans and peas	16.20
Poultry	10.71
Sugar	5.56

Note: Commodity scores sum over domains.

Table 6.7 Commodity-domain rankings

Rank	Commodity	Domain	Score	Rank	Commodity	Domain	Score
1	Sorghum and millet	HLL	54.24	31	Oilseeds	HHH	6.04
2	Oilseeds	HLL	43.19	32	Coffee and tea	LLL	5.97
3	Cassava	HLL	41.55	33	Cassava	HHH	5.87
4	Beef	HLL	31.41	34	Maize	HHH	5.51
5	Milk	HLL	28.00	35	Maize	HLH	4.88
6	Coffee and tea	HLL	25.54	36	Fruits and vegetables	LHH	4.86
7	Maize	HLL	23.09	37	Wheat and barley	HLH	4.38
8	Fruits and vegetables	HLL	20.53	38	Cassava	HLH	4.27
9	Cassava	LLL	20.01	39	Bananas	LHH	3.88
10	Potatoes and sweet potatoes	HLL	19.18	40	Wheat and barley	HHH	3.84
11	Rice	HLL	18.81	41	Coffee and tea	HLH	3.82
12	Bananas	HLL	17.52	42	Coffee and tea	HHH	3.76
13	Beef	LLL	17.22	43	Poultry	HLL	3.73
14	Milk	HHH	17.12	44	Sorghum and millet	HHH	3.48
15	Oilseeds	LLL	15.29	45	Beans and peas	HHH	3.13
16	Milk	LLL	12.79	46	Wheat and barley	LLL	2.92
17	Fruits and vegetables	HHH	12.55	47	Sorghum and millet	HLH	2.90
18	Maize	LLL	11.10	48	Milk	HHL	2.73
19	Sorghum and millet	LLL	10.96	49	Potatoes and sweet potatoes	HHH	2.71
20	Milk	HLH	10.90	50	Sugar	HLL	2.39
21	Bananas	HHH	10.27	51	Poultry	HHH	2.28
22	Fruits and vegetables	LLL	9.38	52	Bananas	LLL	2.13
23	Potatoes and sweet potatoes	LLL	9.24	53	Maize	LHH	2.04
24	Fruits and vegetables	HLH	7.99	54	Beans and peas	LLL	2.02
25	Rice	LLL	7.77	55	Sorghum and millet	HHL	2.01
26	Beans and peas	HLL	7.08	56	Fruits and vegetables	HHL	2.00
27	Milk	LHH	6.63	57	Potatoes and sweet potatoes	HLH	1.97
28	Wheat and barley	HLL	6.48	58	Beans and peas	HLH	1.93
29	Bananas	HLH	6.40	59	Bananas	HHL	1.86
30	Oilseeds	HLH	6.36	60	Beef	HLH	1.73

Notes: Domain refers to agricultural potential/market access/population density—for example, HLL= high potential, low access, low density. H = high; L = low. See text for the method of computing the scores.

fruits and vegetables emerges as second in priority to milk, with bananas, oilseeds, and cassava all of higher priority than maize.

Potential Benefits from Regional Coordination

In theory, regional integration and collective action in agricultural R&D among neighboring countries can lead to economies of scale and spillover benefits that permit research systems to jointly achieve the critical masses and cost savings needed to address problems beyond the capacities of individual systems. Such considerations are far from wishful thinking in ECA. One of the principal mandates of the Association for Strengthening Agricultural Research in Eastern and Central Africa (ASARECA) is to identify, catalyze, and coordinate such activities through its several networks (ASARECA 1997). The networks convene a range of research-based activities aiming to contribute to sustainable economic growth and improved social welfare in ECA while maintaining the quality of the environment. Activities include laboratory-based biochemistry studies, experiment station breeding, farmer participatory technology assessment, commodity market surveys and subsector reviews, and efforts to harmonize regional trade policies. Details on such activities are well documented (for example, EARRNET 2003; ECAPAPA 2003;

Rank	Commodity	Domain	Score	Rank	Commodity	Domain	Score
61	Poultry	LLL	1.71	91	Milk	LHL	0.56
62	Milk	LLH	1.63	92	Sugar	HLH	0.48
63	Oilseeds	LHH	1.61	93	Beef	LHH	0.44
64	Oilseeds	HHL	1.59	94	Rice	LHH	0.43
65	Cassava	LHH	1.51	95	Fruits and vegetables	LHL	0.41
66	Beef	HHH	1.51	96	Sorghum and millet	LLH	0.38
67	Poultry	HLH	1.45	97	Poultry	HHL	0.36
68	Cassava	HHL	1.44	98	Bananas	LLH	0.34
69	Beans and peas	LHH	1.21	99	Cassava	LLH	0.29
70	Maize	HHL	1.21	100	Oilseeds	LHL	0.29
71	Fruits and vegetables	LLH	1.19	101	Beef	LLH	0.28
72	Coffee and tea	HHL	1.16	102	Maize	LHL	0.27
73	Rice	HHH	1.15	103	Cassava	LHL	0.26
74	Sorghum and millet	LHH	1.13	104	Sugar	HHL	0.23
75	Sugar	LLL	1.11	105	Poultry	LLH	0.22
76	Sugar	HHH	1.09	106	Beef	LHL	0.19
77	Beef	HHL	0.97	107	Sorghum and millet	LHL	0.19
78	Oilseeds	LLH	0.96	108	Rice	LHL	0.17
79	Poultry	LHH	0.88	109	Sugar	LHH	0.17
80	Rice	HHL	0.83	110	Bananas	LHL	0.16
81	Rice	HLH	0.83	111	Rice	LLH	0.15
82	Potatoes and sweet potatoes	LHH	0.70	112	Beans and peas	LLH	0.14
83	Potatoes and sweet potatoes	HHL	0.66	113	Potatoes and sweet potatoes	LLH	0.13
84	Coffee and tea	LHH	0.66	114	Potatoes and sweet potatoes	LHL	0.12
85	Wheat and barley	HHL	0.65	115	Wheat and barley	LHL	0.10
86	Beans and peas	HHL	0.64	116	Coffee and tea	LHL	0.08
87	Wheat and barley	LLH	0.64	117	Poultry	LHL	0.07
88	Wheat and barley	LHH	0.59	118	Beans and peas	LHL	0.06
89	Maize	LLH	0.59	119	Sugar	LLH	0.05
90	Coffee and tea	LLH	0.57	120	Sugar	LHL	0.04

Table 6.8 Degree and scope for agricultural research and development spillovers

Commodity	Regional gains without spillovers (US$000s per year) A	Incremental spillover gains (US$000s per year) B	Gain to region from spillovers (%) B/A	Gains outside innovating countries (%) B/(A + B)	Degree of variation of spillover gains outside innovating countries
Cassava	5,200	2,581	50	33.4	2.29
Cow milk	4,456	2,984	67	40.8	1.71
Plantain	6,575	659	10	9.2	2.49
Maize	5,659	1,477	26	20.7	1.99
Beef	3,741	2,409	64	39.2	1.44
Coffee	2,566	1,461	57	37.7	2.22
Sorghum	1,064	2,059	194	66.3	1.83
Vegetables	1,742	956	55	35.4	1.09
Dry beans	1,701	626	37	27.0	1.09
Rice	854	1,355	159	61.3	2.51
Mutton/lamb	467	1,399	300	75.6	1.75
Groundnuts	553	1,254	227	69.5	2.07
Potatoes	982	490	50	33.7	1.32
Cotton	427	251	59	37.1	1.64
Cashew nuts	396	5	1	1.6	3.00
ECA	36,381	19,965			

Source: Abdulai, Diao, and Johnson 2005.

PRAPRACE 2003). At issue here are the potential benefits from regionally coordinated efforts of this kind.[22]

The Dynamic Research Evaluation for Management (DREAM) model is used to quantify the degree and scope of R&D spillovers for such activities for a set of key agricultural commodities in ECA: vegetables, tree nuts, pulses (beans and peas), oil crops, roots and tubers, livestock, fiber crops, and cereals. In applying the model, the major assumptions include (1) technology innovations originate in Kenya, Uganda, and Tanzania (termed the innovating countries); (2) innovations (or any other cost-reducing interventions) are transferable only within ECA—that is, spillovers outside the region are limited by geographic distance; (3) technologies take five years to be fully adopted by farmers at an adoption ceiling of 80 percent; (4) because of imperfect adaptation of technologies between countries, technology spillovers to noninnovating countries are assumed to translate into half the productivity gains initially realized in the innovating countries; and (5) productivity is stimulated by 1 percent increases in each of the three innovating countries and the effect on economic welfare is projected out to 2015 as a stream of annual net benefits. These assumptions are comparable to those in similar studies elsewhere in the region (for example, Mills and Karanja 1997; Kilambya, Nandwa, and Omamo 1998; Mills 1998; Omamo 2002). The results are reported in Table 6.8 and Figures 6.4 and 6.5.

Annual gains from R&D investments that lead to the 1 percent productivity increase total more than US$36 million. Regional spillovers from regionally coordinated initiatives add a further US$20 million to the regional total. Spillover gains range from US$5,000 for cashew nuts to almost US$3 million for milk (Table 6.8, column 2). The results suggest that milk would be an especially big winner from regionally coordinated R&D investments. Not only would benefits to the region from productivity growth in this subsector increase by 70 percent because of spillovers, the benefits

[22]The results reported in this section are drawn from Abdulai, Diao, and Johnson (2005) and Diao et al. (2005).

Figure 6.4 Returns to agricultural research and development investments without spillovers

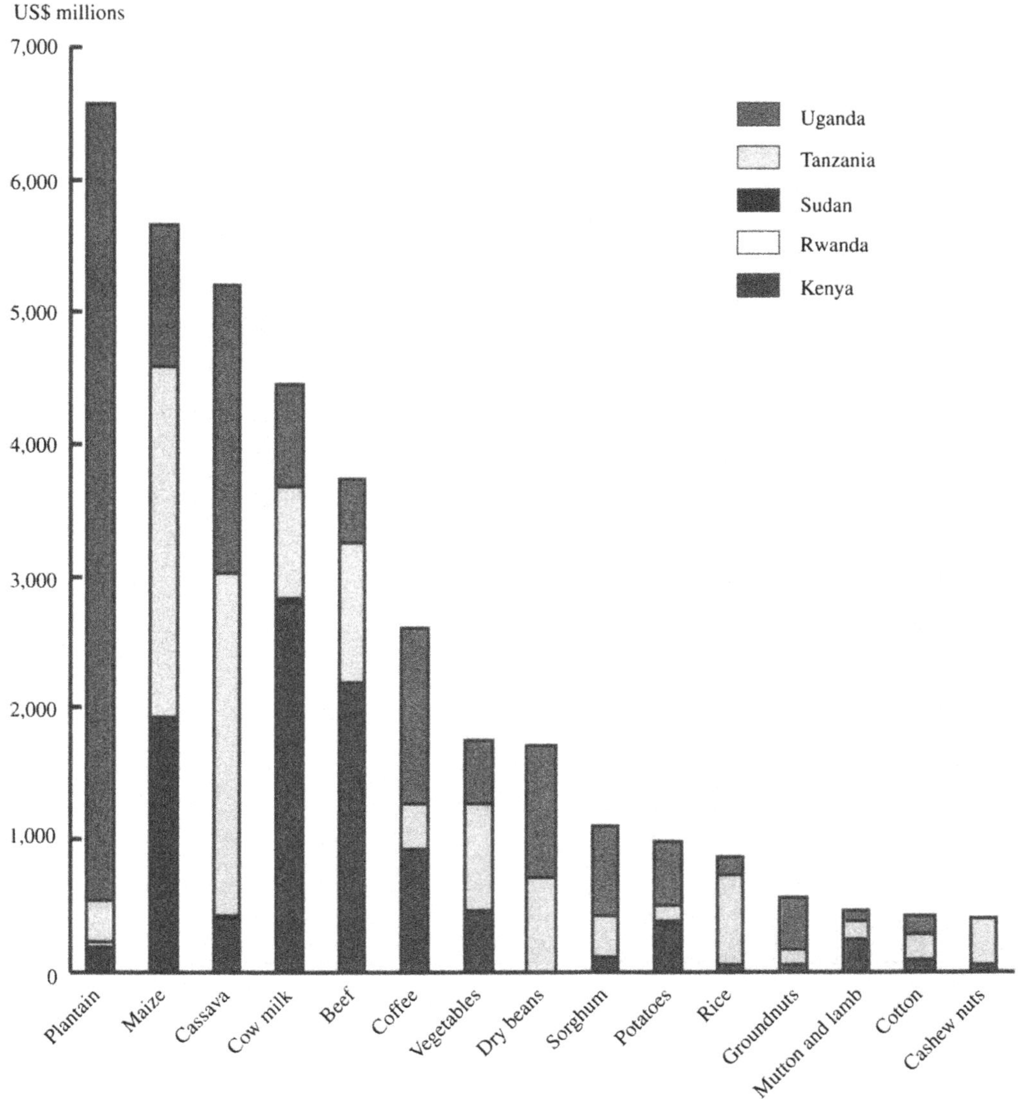

Source: Abdulai, Diao, and Johnson 2005.

accruing to noninnovating countries would be about 40 percent of total regional benefits (Table 6.8, column 4). Productivity growth in the cassava, beef, and sorghum subsectors would also offer high returns from spillovers outside the three innovating countries. Smaller spillover gains would accrue to noninnovating countries for such commodities as plantain, dry beans, potatoes, and maize. These are commodities for which the three innovating countries contribute a large share of the regional totals.

The degree of variation of spillover gains outside the innovating countries (Table 6.8, column 5) captures the equitability in the distribution of these gains across countries. The smaller the degree of variation, the more equitable is the distribution. Commodities exhibiting the most equitable distributions of spillover gains include vegetables, dry beans, and potatoes. Livestock products also seem to offer such scope—dairy, beef, and mutton lamb. The degree of variation of spillover gains would be especially high for cassava,

Figure 6.5 Returns to agricultural research and development investments with spillovers

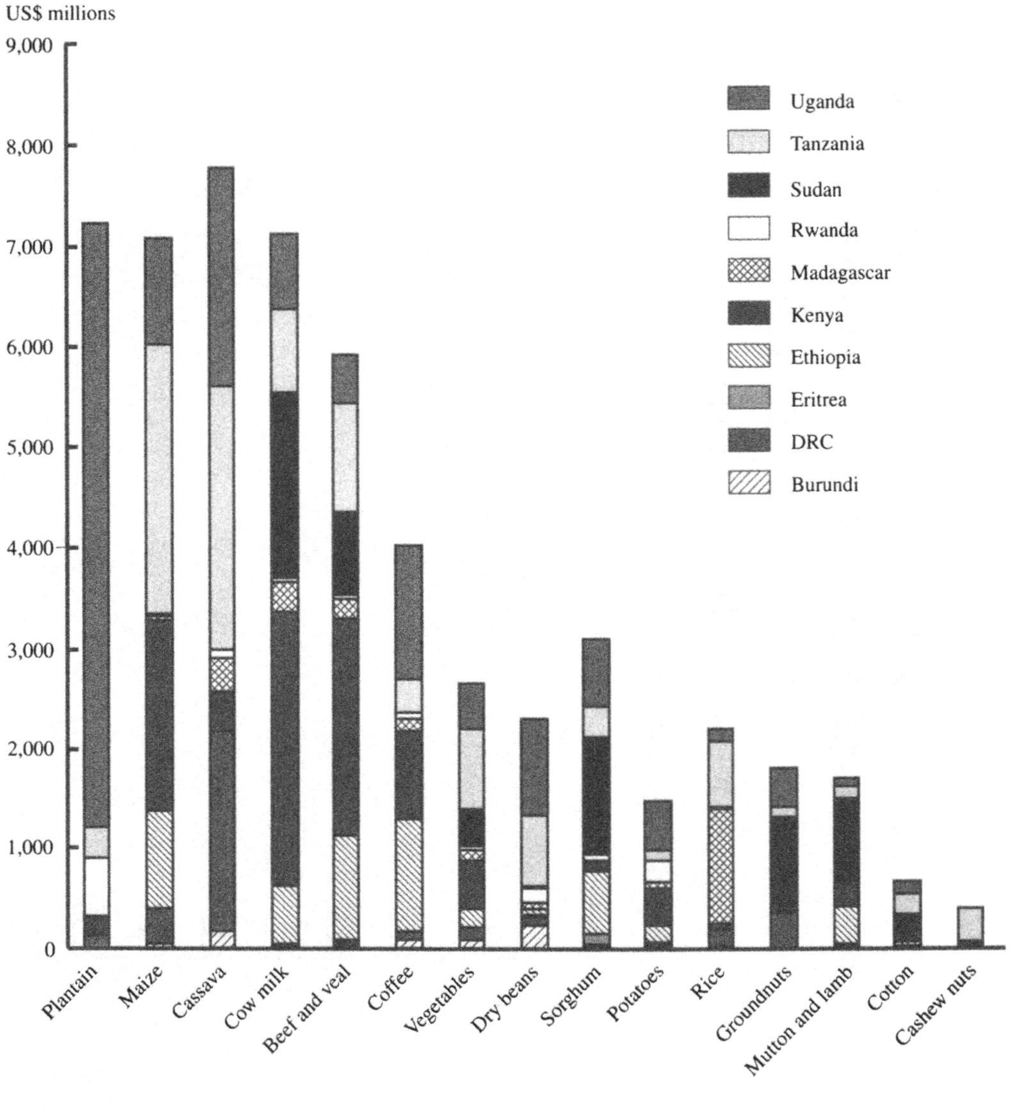

Source: Abdulai, Diao, and Johnson 2005.

plantain, coffee, and rice. As can be seen in Figures 6.4 and 6.5, this disparity is because large shares of the regional spillover benefits from productivity growth in these commodity subsectors would accrue to a small number of countries—to Ethiopia and Sudan for sorghum and mutton/lamb, to DRC for cassava, to Ethiopia for coffee, and to Madagascar for rice. Despite this unevenness in the distribution of the spillover gains, it is clear that most ECA countries stand to gain significantly.

Summary

This chapter outlines strategic priorities for agricultural development in ECA along several dimensions, made possible by an analytical approach that integrates spatial information into an economic modeling framework. When ECA is viewed as a region, milk emerges as the most important commodity subsector for growth-inducing investment in R&D, based on simulated cumulative contributions to overall GDP to 2015. Oilseeds, cassava, and fruits and vegetables also rank highly. Viewed together, the staples subsectors result in the largest GDP gains, followed by livestock products, fruits and vegetables, and oilseeds.

The analysis reveals significant differences between regional and national priorities. Priorities for Kenya, Tanzania, and Uganda closely match those of the region. Regional priorities appear to be less relevant for Eritrea, Ethiopia, and Sudan, with cereals and milk being more important in these countries than in the others. Regional priorities seem to have little relevance for Burundi, DRC, Madagascar, and Rwanda. In Burundi and Rwanda, bananas, potatoes, and sweet potatoes are crucial. In DRC, oilseeds and cassava are paramount. In Madagascar, rice is central.

The analysis also reveals important differences among commodities in the distribution of gains across countries. Whereas growth in the cassava and milk subsectors generates the largest aggregate gains, such gains are concentrated in a handful of countries. Fruits and vegetables, beef, oilseeds, and maize emerge as commodity subsectors in which growth would yield gains that are both large and widespread.

The HLL agricultural development domain emerges as the clear priority for efficient, equitable, and sustainable agriculture-led growth and poverty reduction in ECA. The potential for broad-based benefits from regionally conceived initiatives in agricultural development resides primarily in this domain. That potential appears to be substantial, based on growth in the oilseed, fruit and vegetable, beef, maize, coffee, and tea subsectors. Agriculture-based growth in the LLL, HHH, and HLH domains is also important and likely offers the possibility for both poverty reduction and benefits from regional cooperation. But due to constraints caused by population pressure (HHH and HLH) and biophysical fragility (LLL), such potential is likely to be more difficult to achieve. Given the large number of people residing in these areas, research aimed at overcoming these constraints through improved natural resource management is of high priority.

Agriculture-based growth in the LHH, HHL, LLH, and LHL domains is unlikely to be large enough to warrant major investments in agricultural development. The best growth-enhancing options in these areas probably lie outside agriculture. Any agriculture-related investments in these areas would be justifiable only on equity concerns and only in the absence of more direct avenues for addressing constraints on growth.

The analysis indicates that GDP gains from growth in demand would be equivalent to those from comparable growth in supply. Gains from growth in these two areas would be significantly higher than those from growth caused by disproportionate reductions in barriers to domestic and international trade.

Finally, the analysis identifies significant returns to regional cooperation in agricultural R&D. Regional spillovers from innovations originating in a small subset of ECA countries could equal 40 percent of total regional benefits from agricultural R&D.

CHAPTER 7

Recommendations and Policy Implications

Several recommendations for both national and regional agricultural development policy spring from the analysis presented in the preceding chapters. Clear strategic priorities among agricultural subsectors emerge, as do insights into the nature of agricultural development strategies and policies required to elicit sustained and broad-based growth and poverty reduction. Table 7.1 (similar to Table 3.9 in Chapter 3) summarizes these priorities (and recommendations) and links them to specific agricultural development domains.

Recommendation 1: Spur Productivity Growth in Selected Subsectors

The central implication emerging from the analysis is the importance to sustained growth—in overall gross domestic product (GDP), agricultural GDP (AgGDP), and broad-based poverty reduction—of productivity growth in agricultural subsectors for which there is high and growing demand in the region. Such subsectors include livestock products, major staples, oilseeds, and fruits and vegetables. This is not to minimize the importance of export commodities. Production and marketing of these commodities will remain the most viable options in many areas, most notably in areas with high potential, high market access, and high population density (HHH; see Table 3.8 for the definitions of the agricultural development domains).

Stagnant agricultural productivity growth in eastern and central Africa (ECA) points to the need for increased and more fruitful agricultural research and extension. Advances in science continually create new technological potential. But weaknesses in the region's agricultural research and extension systems prevent translation of that potential into rapid development and wide dissemination and uptake of productivity-enhancing technologies (Chema, Roseboom, and Gilbert 2004). Organizational and institutional challenges appear to be central to overcoming this problem. Traditional approaches to the development and diffusion of agricultural technology are based on centralized, one-way flows of knowledge from the public sector to farmers. Such mechanistic approaches in which public research institutes produce technologies that are disseminated and adopted by farmers have been discredited (Farrington 1998). Interactions among public, private, and collective initiatives are now viewed as crucial to sustainable agricultural innovation and diffusion. A range of options is already being explored in the region, including initiatives that integrate research with outreach and training (for example, NARO 2000), those that seek to privatize agricultural service provision (NAADS 2004), and those that seek to spur linkages among adoption of improved farming practices, farm input supply, and information dissemination (for example, Seward and Okello 2000). Careful monitoring and evaluation of these piloting efforts are crucial, with a view to promoting information

Table 7.1 Agricultural development priorities within ECA development domains

Agricultural potential	Market access	Priorities	Example locations in ECA and potential agricultural development/livelihood options — Population density: High	Low
		Productivity growth Agricultural research and extension systems Weed and pest control Soil and water management Awareness raising and consensus building on biotechnology-related opportunities and risks	*Example locations:* Parts of central and western Kenya, Uganda's Lake Victoria Crescent, parts of central and southwestern and southeastern highlands of Ethiopia, parts of Rwanda and Burundi	*Example locations:* Isolated areas scattered throughout region
	High	*Market improvement* Market intelligence (domestic, regional and international) *Linkages with nonagriculture* Storage, processing, distribution Agro-industrialization	*Options* High-input cereals (for example, maize, rice, wheat) Perishable cash crops (for example, vegetables, fruits, flowers, ornamentals) Intensive livestock (for example, dairy, chickens, pigs) Nonperishable cash crops (for example, coffee, tea)	*Options* As for high population density plus more extensive high-value options (for example cotton, tea, oil crops, fruits)
High		*Productivity growth* Agricultural research and extension systems Weed and pest control Soil and water management Awareness raising and consensus building on biotechnology-related opportunities and risks	*Example locations:* Southwestern Uganda, parts of central and western Kenya, much of the Ethiopian highlands, northern Tanzania, Rwanda, and Burundi	*Example locations:* Large areas of all countries: most of central DRC, southern Sudan, parts of central Uganda, Kenya, and Tanzania, widely scattered areas in Ethiopia and Madagascar
		Market improvement Market development (infrastructure, market information systems, credit institutions, and the like) *Linkages with nonagriculture* Storage, processing, distribution	*Options* High-input cereals (for example, maize, rice, wheat) Nonperishable cash crops	*Options* Intensification in nonperishable crops (cereals, oilseeds, tea, coffee) Livestock intensification; improved grazing areas
		Productivity growth Agricultural research and extension systems Weed and pest control Soil and water management Awareness raising and consensus building on biotechnology-related opportunities and risks	*Example locations:* Parts of northern Ethiopia and central Eritrea, north-central Sudan, western Kenya, Rwanda, and Burundi	*Example locations:* Isolated areas scattered throughout region

(continued)

Table 7.1—Continued

Agricultural potential	Market access	Priorities	Example locations in ECA and potential agricultural development/livelihood options — Population density: High	Low
		Irrigation *Market improvement* Market intelligence (domestic, regional, international) *Linkages with nonagriculture* Storage, processing, distribution	*Options* With irrigation investment High-input cereals Perishable cash crops Dairy, intensive livestock Without irrigation investment Low-input cereals	*Options* With irrigation investment High-input cereals Perishable cash crops Dairy, intensive livestock Without irrigation investment Low-input cereals Livestock intensification, improved grazing areas Woodlots
		Productivity growth Agricultural research and extension systems Weed and pest control Soil and water management Raising awareness and building consensus on biotechnology-related opportunities and risks	*Example locations:* Northern and eastern highlands of Ethiopia, parts of Western Kenya, Rwanda, eastern DRC near lakes Edward and Kivu	*Example locations:* Some lowland areas in Ethiopia and Eritrea, central Sudan, southeastern and northern Kenya, eastern DRC
		Market improvement Market development (infrastructure, market information systems, credit institutions, and the like) *Linkages with nonagriculture* Storage, processing, distribution	*Options* Low-input cereals Limited livestock intensification Emigration	*Options* Low-input cereals Livestock intensification, improved pasture management, improved nutrition, breeding for disease resistance

Source: Compiled by authors drawing extensively on empirical research from ECA, especially in Uganda, Ethiopia, and Kenya (in particular, see Pender, Place, and Ehui 1999; Pender 2004; Ehui and Pender 2005).

sharing, avoiding wasteful repetition of mistakes, and scaling up successes within and across countries.

Specific opportunities for action are likely to vary by agricultural development domain, and thus by the commodities produced within those domains. In the highest-priority domain, HLL, where agricultural potential is high but infrastructure poor, technical change toward more intensive production of key nonperishable commodities is paramount. Staples are likely to continue to feature prominently, as are oilseeds and such traditional cash crops as coffee and tea. As shown in Chapter 6, productivity growth in livestock systems would be extremely rewarding in these areas, but would need to be approached carefully, given the perishability of livestock products.

The emergence of milk presents major challenges to ECA. Kenya is the only country in the region that has succeeded in tapping the potential of its dairy industry (Owango et al. 1998; Ngigi 2004). Despite considerable effort in upgrading cattle and feed resources, countries like Ethiopia and Uganda have failed to generate dynamic growth in milk production. The Kenyan experience suggests that significant private sector investment is required to expand milk marketing infrastructure (including cooling tanks, collection centers, and transportation). Also key is smallholder ability to purchase improved breeds of cattle; for example, Holloway and Ehui (2001) show that farmers in the Ethiopian highlands need on average about three crossbred cows to enter the milk market. Investments to ensure nutritional gains and moderation of lean-season malnutrition are also important. And finally, the regulatory framework must promote market growth, especially the raw milk segment, building on the indigenous market to the extent possible, as indigenous markets are typically highly labor intensive and are known to generate greater levels of employment (Staal and Kaguongo 2003). To realize the huge potential of the dairy subsector in the region, other countries must pay close attention to this set of issues.

As noted in Chapter 5, the relatively high potential and low population density in HLL agricultural domain implies a less sharp trade-off between growth and sustainability in these zones than in the LLL, HHH, and HLH domains that also emerge as key. In the LLL areas, livestock systems undergird livelihoods. Improved animal health, breeding for disease resistance and improved nutrition, and improved pasture management are therefore crucial. Challenges in natural resource management are raised by the extensive nature of most livestock production systems in these areas and thus by the need for collective management of key resources. Development of property rights regimes that improve long-term incentives to invest in, sustain, and improve these resources is therefore important. Where irrigation is an option and cropping feasible, similar institutional challenges are raised by the likely need for community management of irrigation schemes. In such cases, new institutional arrangements will probably be required to develop new sets of rules governing claims to water resources and to determine the means by which water resources are managed by individuals or groups.

In HHH and HLH zones, natural ecosystems are under pressure from persistently low levels of agricultural productivity growth coupled with growing food demand. Productivity growth will hinge on solutions to such problems as soil nutrient depletion, soil erosion, pests, and weeds. Many of the technologies required to address these problems already exist. Most of these technologies are knowledge-intensive, implying the need for structures and processes that promote sustained learning among both farmers and those service providers likely to be involved in successful technology adoption. The difficulties raised by soil nutrient depletion in these areas cannot be overstated. Again, sustainable solutions will probably hinge on institutional innovations—those that lead to efficient and sustainable demand and supply of fertilizers and improved seeds, along with shifts in production practices toward greater commercialization.

The promise of modern biotechnology in spurring productivity growth in the region remains only partially exploited.[23] Although that promise cannot be ignored, the role of this branch of science in the economic transformation and sustainable development of ECA (and other parts of Africa) is subject to increasing debate and controversy. Two extreme positions polarize the debate. The extreme pro-biotechnology groups catalog the potential benefits of biotechnology (for example, enhanced taste and quality of foods, nutritionally enhanced foods for chronically malnourished populations, reduced maturation times for crops leading to labor savings, enhanced tolerance to biotic and abiotic stresses for crops leading to reduced dependence on herbicides and

[23]A range of conventional biotechnology-related activities (such as tissue culture and marker-assisted selection) are under way in several countries in the region. But only in Kenya have field-based transformation events been recorded (Sengooba 2006). Governments have established some interim structures to serve as coordinating and advisory bodies (as well as for enforcement of biosafety regulations), but comprehensive parliamentary bills addressing all aspects of biotechnology and biosafety have yet to be developed.

pesticides, and enhanced disease resistance in livestock leading to reduced dependence on drugs) and often dismiss any concerns about potential risks. They tend to portray biotechnology as the ultimate panacea for food insecurity in ECA. At the other extreme are the anti-biotechnology activists, who see no evident benefits, associate all biotechnology with genetically modified organisms (GMOs), and therefore link biotechnology with nothing but danger and risks (such as unscrupulous profiteering by GMO-producing private firms and negative effects on the environment and human health of release of development genetically modified varieties and foods). They demand that the development, commercialization, and application of biotechnologies be stopped. The stark contrast between these two views has left many ECA policymakers and sections of the public uncertain about how to proceed, because reliable information and guidance are lacking. Increasing uncertainty and confusion is evident in the responses of many ECA governments to a wide range of social, ethical, environmental, trade, and economic issues associated with the development and application of modern biotechnology. This trend could deny ECA citizens the opportunities to derive benefits from biotechnology while minimizing the associated risks. ECA countries need to be in a position to make informed choices and establish policies and strategies to judiciously respond to developments associated with biotechnology—for example, by developing appropriate biosafety procedures and intellectual property rights. ECA policymakers and citizens must therefore seize the biotechnology agenda for themselves. To do so requires a greater clarity in concepts, facts, and options for establishing consistent institutions and policies governing biotechnology in ECA agriculture and food security. Specifically, there is a pressing need to raise awareness, promote dialogue, and catalyze consensus-building mechanisms among national and regional stakeholder groups spanning public bodies (including parliamentary and judicial organs), the private sector, and civil society. Support for ongoing efforts with such aims is crucial (for example, APDB 2004; RABESA 2004).

Recommendation 2: Strengthen Agricultural Markets

The analysis confirms that productivity growth without significant improvements in market functioning is counterproductive. Physical impediments to agricultural trade and exchange related to poor infrastructure remain as high in ECA as they are elsewhere in Africa (Pederson 2000). Such impediments point to the need for major investments in roads, railways, and telecommunications. Absent such investments, the scope for sustained agricultural productivity growth in ECA will remain extremely limited.

A very real danger is that today's HLL zones—with all their promise—could become tomorrow's HLH zones—with all their problems stemming from high agricultural potential but high population density and low market access. To avoid such an outcome as rural populations grow, major investments in rural infrastructure are urgently required. Such a conclusion is unremarkable, but the analysis allows greater specificity than is usually the case. If ECA governments expect such investments to spur broad-based growth and poverty reduction, then they should give priority to HLL areas, aiming to better connect these zones with locales where demand for produce is growing rapidly. Such locales include not only urban areas but densely populated rural regions—for example, the HHH and HLH agricultural development domains.

But agricultural markets in ECA are fraught with constraints and inefficiencies that are not always linked to poor infrastructure. Access to credit is limited for most agricultural traders. Costs are high for obtaining market information, searching for buyers and sellers, and enforcing contracts. Agricultural trade is risky, personalized, and cash-based, with limited long-term investment by private traders in transport or storage, even in regions with relatively good infrastructure. Limited storage capacity and poor access to formal financing mechanisms render prices highly volatile. Other important institutional constraints include a general lack of adequate market information, lack of grades and standards, low levels of market transparency, and frail legal environments governing property rights and contract enforcement (Gabre-Madhin 2001). These constraints imply several noncompetitive elements in agricultural markets, especially in smallholder-dominated areas.

Despite their ubiquity, these institutional constraints are context specific. For instance, market development for cash crops is a totally different (and easier) proposition than it is for staple food crops (Jones 1972; Pearson, Timmer, and Falcon 1983; Poulton et al. 2004). Incentives for private sector investment in market infrastructure on the scale required to pull in large volumes of commodities are higher for cash crops with high unit returns and well-defined and relatively narrow marketing channels than they are for staples, which are often produced and marketed in small quantities over large areas. Again, the principal challenges are organizational. But at present there are more questions in this regard than there are answers.

How can entry by traders into smallholder areas be induced at levels sufficient to invoke the economies of scale needed for broad-based economic development? Which institutional innovations in agricultural markets are required? Which of these innovations can feasibly be left to traders and local communities, and which ones constitute residual roles for public sectors? How can the imperfect competition likely to be inherent in new markets be addressed? Which market organizations support rapid dissemination of information without compromising behavior that is individually costly but beneficial when reciprocated (for example, farmer uptake of improved technology and trader commitment to provide related inputs or purchase resulting outputs)? What is the scope for farmer collectives (such as community groups and associations) to inspire procompetitive market outcomes through collective bargaining with traders and processors?

Answers to some of these questions are beginning to emerge as experiments with alternative institutional arrangements are undertaken. Emerging solutions include new smallholder-oriented market information systems and commodity exchanges (KACE 2005); efforts to reduce transaction costs, enhance quality, and access new markets through smallholder collective action (SACRED-Africa 2005); and subsector-specific public support and regulation of private sector activity in agricultural markets (Poulton et al. 2004). Once again, careful monitoring and evaluation of these efforts to promote information sharing, learning, and scaling up within and across countries is crucial.

Just as onerous state domination of markets revealed the limits of the public sector in promoting market development, so, too, did blind liberalization of markets uncover the absence of several crucial market development capacities in the private sector. The task of realizing the promise of markets in agricultural and broader economic development is clearly too important to leave wholly in the hands of either sector. A division of labor that yielded broad-based benefits in other parts of the world (Sabel 1994) would involve governments supplying those forms of market stabilization that only governments can provide (such as loan and credit guarantees), but only in return for commitments from the private sector to invest in specified institutional arrangements and organizational forms (for example, establishing input supply networks in smallholder areas). Specifying the desired arrangements and forms in given subsectors would therefore be logical first steps for governments, followed by measures that catalyze, facilitate, and monitor their evolution in the private sector (Sabel 1994).

Recommendation 3: Promote Growth Linkages

The analysis points to the importance of growth in nonagricultural sectors to sustained growth in the agricultural sector. Such growth in nonagricultural sectors not only provides crucial off-farm employment and income opportunities for rural populations, it also generates demand for agricultural products (Delgado et al. 1998). These effects cut in both directions. By increasing effective demand for goods and services of a broad base of rural dwellers, agricultural growth benefits that are widespread can be effective at capturing growth opportunities offered by linkages to nonagricultural sectors. Targeted investment to promote linkages with key nonagricultural sectors and spur growth in those sectors is therefore crucial.

In areas where transport costs and other structural factors isolate local economies from outside sources of demand for local products, the strongest links between agricultural and nonagricultural sectors spring from production and consumption of nontradable commodities. This observation points to a role for measures that both improve production and promote demand for income-generating enhancements in local produce through improvements in storage, processing, and distribution. The emergence of the low market access HLL and HLH domains as high priorities for growth and poverty reduction suggests that this type of measure should also be a priority. Milk and oilseeds appear to be especially promising target commodities. Their markets are large and their links to agroindustrialization immediate. Artisanal processing is often also possible. Large-scale processors may have incentives for vertical coordination with producers through contract farming schemes, thereby tackling issues of support for service provision.

In areas where tradables are more important and the supply-side is the clear constraint on growth (such as the HHH domain), policies that promote agroindustrialization will be key. Three related sets of measures would be appropriate: growth of agroprocessing, distribution, and farm-input provision off-farm; institutional and organizational adjustments in relations among agroindustrial firms and farms (including greater vertical integration); and concomitant changes in product composition, technology, and sectoral and market structures.

No fully generalizable prescription exists for catalyzing and strengthening growth linkages in all instances. However, Haggblade, Hazell, and Reardon (2002) suggest three broad principles that permit identification of cost-effective interventions across a broad diversity of specific settings.[24] First, development agencies should identify key engines of growth. The current analysis sheds light on the crucial aspects of such engines for ECA countries. Second, to facilitate the systematic search for cost-effective interventions by identifying large numbers of like firms facing similar opportunities or constraints, development agencies should focus on subsector-specific supply chains. This emphasis should provide a tractable means for prioritizing key infrastructure requirements and tracing commodity flows across space, ensure a focus on final markets, and enforce the necessary link between evolving consumer requirements and supply systems. It also highlights competitive and complementary relationships among firms of different sizes and underscores specific opportunities and threats confronting the rural poor. Third, development agencies must build flexible institutional coalitions. Rather than creating expensive new integrated bureaucracies, stakeholders must find ways to

[24]Similar proposals are found in Tomich, Kilby, and Johnston (1995) and Lanjouw and Feder (2001).

work across the existing patchwork of private and public agencies that currently exist. Depending on the commodity subsectors selected for review, a coalition of key stakeholders may include government regulators, technical institutes, industry associations, key private sector participants, donors, or nongovernmental organizations.

Recommendation 4: Exploit Opportunities for Regional Cooperation

The analysis identifies significant returns to regionally coordinated initiatives in agricultural development. Each of the above recommendations has a regional dimension.

Many crops are grown throughout the region, many in the high priority HLL domain prominent in most countries. As illustrated in the previous chapter, opportunities to develop regionally conceived and implemented agricultural research and development (R&D) initiatives must be exploited. Spillover benefits from such efforts are likely to be large and must be grasped, especially in the expansive HLL domain. Development of regional biosafety procedures and intellectual property rights may help the region avoid wasteful duplication of effort as it struggles to come to terms with the challenges posed by biotechnologies.

Most of the abovementioned market constraints occur throughout ECA. Solutions identified in given countries may apply in others. Opportunities to reduce learning costs through deliberate information sharing should be seized. Where market constraints are linked to poor infrastructure, high returns to regional initiatives to improve rural infrastructure may exist. Obvious examples include improvement and harmonization of the region's telecommunications and transport systems and removal of policy-related barriers to movement of goods across borders.

Cross-sectoral linkages have both national and regional manifestations. For instance, measures that simultaneously target sustained growth in plant husbandry, on one hand, and in agroprocessing, on the other, might require national action to ensure expanded farmer and trader access to key technologies and information, and regional action to standardize grades and quality requirements.

The challenges to realizing such benefits from regional cooperation are political, institutional, and organizational. As noted at the beginning of this report, there are at present no regional bodies with mandates to coordinate and implement agricultural development policy in ECA. However, even without such bodies, several regional agricultural development initiatives have been initiated and implemented with considerable success (for example, ECAPAPA 2002; RATES 2005; RATIN 2005). Given the high-level attention currently being accorded regional cooperation in ECA and elsewhere in Africa, it is worth reflecting on the details of one of these initiatives, namely, the effort by the Eastern and Central Africa Program for Agricultural Policy Analysis (ECAPAPA) to convene a process leading to harmonization and rationalization of seed policies and regulations in ECA (ECAPAPA 2002).

This relatively modest initiative has yielded concrete trade-enhancing outcomes.[25] Research findings coupled with focused deliberations among stakeholders drawn from the public and private sectors led to agreements that reduced the number of quarantined pests from 33 to 3, thereby reducing the period required for issuance of phytosanitary certificates from two weeks to two days, with major cost savings. Participants agreed on uniform procedures for variety evaluation, reducing the number of seasons required for National Variety Performance

[25]The initiative costs an average of US$140,000 per year and covers activities in Burundi, Democratic Republic of Congo (DRC), Eritrea, Ethiopia, Kenya, Madagascar, Rwanda, Sudan, Tanzania, and Uganda—that is, about US$14,000 per country per year.

Trials from six to two. Major rationalization of requirements for seed imports and exports was achieved in all participating countries, leading in some cases to a one-stop, one-person process, whereas in the past several people in distant locales would have been involved. National and regional seed trader associations have been created, with much of the impetus coming from private sector participants. Significantly, these activities have taken place consistently over five years—a period during which some participating countries have been at war and interaction among relevant public agencies has been limited. Perhaps more important, the initiative has created a new forum for engagement for representatives of a wide range of stakeholder groups, resulting in major improvements in welfare. As a model for future efforts in focused, nimble, low-cost, and nonpolitical but effective regional cooperation in agricultural development, this is an extremely informative initiative.

CHAPTER 8

Conclusions

The central problem facing agricultural and general economic policymakers in eastern and central Africa (ECA) today is how to promote self-sustaining processes of growth and poverty reduction fueled by technological advances in small-scale agricultural production and trade. This report has put forward a set of strategic priorities for agricultural development in the region with that problem in view. In some respects, the stated priorities and recommendations are not surprising. Hardly anyone would argue that it is not important that ECA countries foster productivity growth, strengthen markets, improve linkages between agricultural and nonagricultural sectors in rural areas, and promote regional cooperation. But the analysis also yields findings that are not obvious and certainly not widely appreciated.

First, the analysis indicates that the greatest potential for agriculture-led growth and poverty reduction in the region lies in agricultural subsectors serving domestic and regional markets—not those directed at overseas markets. Export commodities will continue to be crucial income earners in key parts of ECA, but they will not be the answer to the problem of widespread poverty and hunger in the region. Second, the analysis indicates that among agricultural subsectors for which there is large and growing domestic and regional demand, staples loom large as a group. Production and sale of these "poor man" crops can be pathways out of poverty for millions of citizens of ECA.

Third, because growth in those areas with high agricultural potential, low market access, and low population density (the HLL domain) most powerfully affects gross domestic product (GDP) and agricultural GDP, overall growth in most ECA countries could be enhanced and poverty reduced without major trade-offs with environmental sustainability. Such "win-win-win" solutions are rare. They are also ephemeral. This opportunity for sustainable growth and poverty reduction for individual ECA countries and the region as a whole is unlikely to last for long. It must be grasped immediately.

But then the question is how? How can ECA countries respond meaningfully to these priorities, either individually or in tandem? The institutional and political environment within which agricultural policy is formulated and implemented is complex. As noted above, agriculture has numerous linkages with other sectors. Agricultural policy formulation and implementation therefore face unique cross-sectoral (horizontal) demands. Political tensions emerge as agriculture's line ministries negotiate terms with one another, with the executive branch, and with various nonline ministries that influence resource allocation for national development (such as ministries of finance and planning).

Concerns for such horizontal perspectives in agricultural policy formulation and implementation have been implicit throughout this report. Such concerns underpinned the study's overarching motivation; they guided the choice of analytical approaches and drove the analysis itself. This study's answer to the question of "how" has therefore been driven by two crucial—

but typically unaddressed—questions of "what": What can agriculture do for the rest of the economy if given a real chance? What needs to be done within agriculture to achieve results? The answers are important and groundbreaking, but they are not enough. Policymakers want to know how to accomplish these tasks. But real how-answers will never emerge from reports like this one, no matter how comprehensive or enlightening they might be. Real answers to how to promote growth-enhancing, poverty-reducing agricultural development in ECA will emerge only as countries come to grips with the strategic priorities they face in agricultural development, align resource allocations with those priorities, and, perhaps most crucially, fashion new institutional arrangements and processes that translate the outputs of hard-working ECA citizens into tangible and sizable private benefits. The challenge facing the region is enormous. This study suggests that so, too, are the potential rewards.

APPENDIX A

Spatial Perspectives of Development Strategy Formulation and Implementation

Formulating and evaluating agricultural development strategies for a region as large and diverse as eastern and central Africa (ECA) is difficult. The task must be decomposed in a number of ways. In the main text we described one such method, namely, to dissect the region into geographical units—dubbed development domains—in which similar agricultural development problems or opportunities[26] are likely to occur. In devising such an approach, we recognized the difficult trade-off faced between defining (1) a small enough number of domains to meet the constraints of pragmatism, policy communication, and available data and (2) a sufficiently large number of domains to properly reflect the rich variety of socioeconomic and biophysical conditions that exist in ECA.

Regardless of the number, however, a key goal has been to use a single set of domain criteria and to apply them consistently across the region. Only with such a consistent approach can the true similarity or dissimilarity of conditions existing in, say, the highlands of Tigray in Ethiopia, be properly compared and contrasted with locations in other ECA countries. If we can consistently group locations throughout the region in accordance to their similarity, we are well on the way to addressing some key strategic development issues:

- Where are those geographic areas within and across countries in ECA in which development problems and opportunities are likely to be most similar?
- Where will specific types of development policies, investments, and incentives be most cost effective?
- Given the lessons of successful practices from one location, where else in ECA might those lessons be applied? Such areas might be targets for the replication of these successes.[27]

We have implemented this approach by using mapped information in a geographical information system (GIS). A GIS is a computer-based environment in which data with spatial properties can be represented and analyzed. A grid-based GIS allows representation of variation

[26]From a national and regional strategic and policy perspective, development domains represent areas of broadly similar strategic and investment opportunities. From a farm or enterprise perspective, development domains offer a way of identifying viable sets of livelihood options.

[27]The source location may even lie outside ECA. For instance, new maize or soybean varieties being used in certain agroecological zones of Brazil might have potential for application in similar zones in Sub-Saharan Africa. Another example might be a biotechnology tool available from an advanced research institute in Asia that could be applied by researchers in Sub-Saharan Africa.

in attributes of interest (for example, annual rainfall, population density) across a grid that covers the entire ECA region. The grid is notionally of any size, but typically for a region the size of ECA, the individual grids (pixels) of information range from 1 × 1 to 20 × 20 kilometers (1 to 400 square kilometers) in size. Thus ECA would be represented by more than 8 million and 20,000 pixels, respectively. Each pixel corresponds to a specific location and can be linked to a stack of attribute values derived from different thematic regional grids. GIS then provides a framework for looking at local patterns of association among attributes and examining how these patterns vary over broad areas. In the present context, we are most interested in the geographies of attributes that constrain or enable different agricultural development options. Efforts to promote sustainable agricultural development are more likely to be effective if domains having similar attributes can be reliably delineated. Through appropriate regional stratification, more focused evaluations might be made of those development strategies that best exploit the comparative advantage of individual locales.

Development Domains for ECA

In this section we briefly review previous work on defining development domains in ECA and discuss the outcome of technical consultations on updating this work. We then describe the enhanced approach and the results of the development domain analysis for ECA. We also illustrate the linkages between individual development domains and specific development strategies in the region.

Past Work

Empirical evidence from Sub-Saharan Africa and beyond suggests that diverse development strategies are needed. Some policy and strategy components are of general relevance: improved security; political and macroeconomic stability; and adequate public investments in basic health, education, and physical infrastructure. But designing development strategies to support specific economic opportunities, and especially those related to agriculture, relies on an understanding of the comparative advantage of individual locations (Pender, Place, and Ehui 1999; Pender et al. 2004). Thus from a strategy formulation perspective, a high priority is to recognize differences in comparative advantage for agriculture-based livelihood opportunities across ECA.

Examples taken from the analysis of livelihood strategies in Uganda and Ethiopia show, for instance, that development of such high-value perishable commodities as horticultural crops or dairy has been greatest in areas with relatively high market access and agricultural potential. In such areas, investments in appropriate forms of infrastructure, human capital, and institutions appear to have yielded higher social returns and facilitated sustainable agricultural development. In areas more remote from markets or having lower agricultural potential, alternative income strategies, such as extensive livestock production or forestry activities, appear to have greater comparative advantage (Pender 2004). In Uganda, those areas that exhibited medium-to-high agricultural production potential, low-to-medium population density and low-to-medium market access often engaged in the expansion of food staples: cassava, beans, and maize (Pender et al. 2004).

Drawing on early findings of this type from Central America, Ethiopia, and Uganda, Pender, Place, and Ehui (1999) proposed the notion of a development domain as a geographical area or set of geographical areas endowed with similar comparative advantages, based on similar agricultural potential, access to markets, and population density, and subsequently considered the relevance of such domains for ECA. These researchers developed a typology that assigned high or low status to each of the above three factors and that, in combination, defined eight distinct development domains for ECA.[28] For each domain, they identified broad geographic subregions of ECA where different combinations of factor values were found, as summarized in Table 3.8 (for example, high-high-high [HHH] reflects conditions in central Kenya and low-low-low [LLL] is typical of much of northern Ethiopia). They also linked each domain to conditions for which development strategies that have been—or are hypothesized to be—successful (for example, HHH is favorable to high-input cereals and perishable cash crops, and LLL to low-input cereals and livestock intensification). This early work provided no spatial representation of development domains, in part because the specific measures of high and low agricultural potential, market access, and population density had not been defined. Wood et al. (1999) extended this work by both defining specific metrics (for example, length of growing period to represent agricultural potential) and presenting those metrics in gridded map form. By overlaying (intersecting) the three data grids in a GIS, a first set of mapped development domains was generated for a large share of the ECA region.

As summarized by Pender et al. (2004, 769):

> Agricultural potential largely influences the absolute advantage (productivity) of a location in production of particular agricultural commodities, while access to markets and infrastructure and population pressure help to determine the comparative advantage (profitability) of particular livelihoods, given the absolute advantages (Pender, Place, & Ehui, 1999). For example, an area with suitable climate and soils may have an absolute advantage in producing high-value perishable vegetables, but little comparative advantage in this if it is remote from markets and roads. Improvements in market or road access are thus expected to favor production of higher value perishable commodities as well as non-farm activities, and should contribute to higher incomes and welfare (Pender, Scherr, & Duron, 2001a). Improved access to markets and infrastructure has more ambiguous theoretical impacts on land use, land management practices and resource conditions, depending upon the relative impacts on costs of productive factors (Angelsen, 1999; Pender et al., 2001a), and because of ambiguous effects of output prices on incentives to conserve land (LaFrance, 1992; Pagiola, 1996). Population density is expected to influence the labor intensity of agricultural production, including the choice of commodities as well as production technologies and land management practices, by affecting the land-labor ratio (Boserup, 1965; Pender, 2001). Population growth may drive expansion of agricultural production into forest or grazing areas, reduction in fallow, or induce adoption of land-saving commodities or technologies, investments in land improvement, and adoption of labor-intensive land management practices, among other changes (Pender, 2001; Tiffen, Mortimore, & Gichuki, 1994). Without improvements in technologies, markets or infrastructure, population-induced intensification is unlikely to improve welfare, though it may improve resource conditions by inducing land conservation (Pender, 2001; Tiffen et al., 1994).

Reviewing and Extending the Basic Approach

In preparing for the current study, the previous work was first reviewed. In addition to the need to extend its geographic coverage to include 10 ECA countries, it was decided

[28]Pender, Place, and Ehui (1999) focused primarily on interpreting the domain scheme for Ethiopia, Kenya, and Uganda. The subsequent mapping by Wood et al. (1999) included Burundi, Rwanda, and northern Tanzania.

to reexamine the definition of development domains in several ways. The first was to confirm the continued validity of the three defining factors—agricultural potential, market access, and population density—in this broader geographical context. The second was to review the methods by which each factor was determined, and the third was to obtain the most recent and reliable data to support the analysis. To help achieve these goals, two technical consultations were held (in Nairobi and Kampala in July 2004).

Review of the most recent empirical work from Uganda, Ethiopia, and Kenya (Pender et al. 2004; Pender, Place, and Ehui 2006), coupled with expert discussion at the technical consultations, reaffirmed that agricultural potential, market access, and population density have good explanatory power. They can help distinguish those communities and geographic areas where specific agriculture-related livelihood opportunities are more likely to occur and be economically beneficial. One important finding was that a range of variables might be used to proxy each of the individual factors. For example, in their Ugandan study, Pender et al. (2004) used a compound agroclimatic variable—not just length of growing period—as their metric of (rainfed) agricultural potential, and that variable contained information on rainfall amount and seasonality, length of growing period, and February temperature. Similarly, different metrics of market access have been used, ranging from a simple "distance to market" variable to a compound market integration index that includes a measure of the time of travel (taking account of the relative difficulty of travel on different classes of pathways and roads) and of market size. Reaching agreement on which specific variables were both most relevant and feasible to represent at the regional scale was the subject of expert consultation.

The technical consultations held in Nairobi and Kampala therefore addressed two key issues and generated additional recommendations. The first was to assess what specific issues and minimum amount of information must be embodied in the metrics of agricultural potential, market access, and population density so as to assess their variation across ECA in meaningful ways. The second was to consider ways of further enriching the development domain concepts and practices in the light of the specific needs of the ECA studies. These were (1) to consider how domains could be defined and used in a hierarchical way to match increasing levels of specificity in development strategy formulation and (2) how to link development domain analysis to that of the Association for Strengthening Agricultural Research in Eastern and Central Africa (ASARECA) Natural Resource Management Group. In this report, we focus on the method and data issues for describing the three key factors and development domains at a strategic level, but it is worthwhile to consider even at this stage the relevance and feasibility of generating more operationally oriented domains as strategy formulation, investment planning, and implementation proceed.

We illustrate this procedure in Box A.1, which highlights how to maintain a single set of generic domains for all levels of regional treatment, with indications of how that might fail to capitalize on opportunities to better match the known favorable conditions for specific options as those options become better articulated. In general, as strategies become better defined, it makes sense to fine-tune domain definitions to better delineate the most promising target intervention areas. The generation and use of these nested domains could mirror organizational and thematic mandates.

Agricultural Potential

The key factor that endows any location with a comparative advantage to support agriculture-based livelihoods is its biophysical potential to nurture the growth of economically important plants and animals. Location-specific factors strongly influence the types of crop and livestock that will perform well, the risk of exposure to harmful

Box A.1 Hierarchical Approach to the Definition and Application of Development Domains in ECA

One goal of the spatial analysis supporting the ECA strategic studies is to provide a flexible framework for the definition and use of development domains. Much of the discussion and analysis of this report is focused on the generation of a single, implicitly generic development domain map of ECA that helps delineate broad spatial patterns in the likely comparative advantage of different subsectoral strategies across the region. This span of strategic focus is depicted by the highest diamond in the figure. At this level the intent

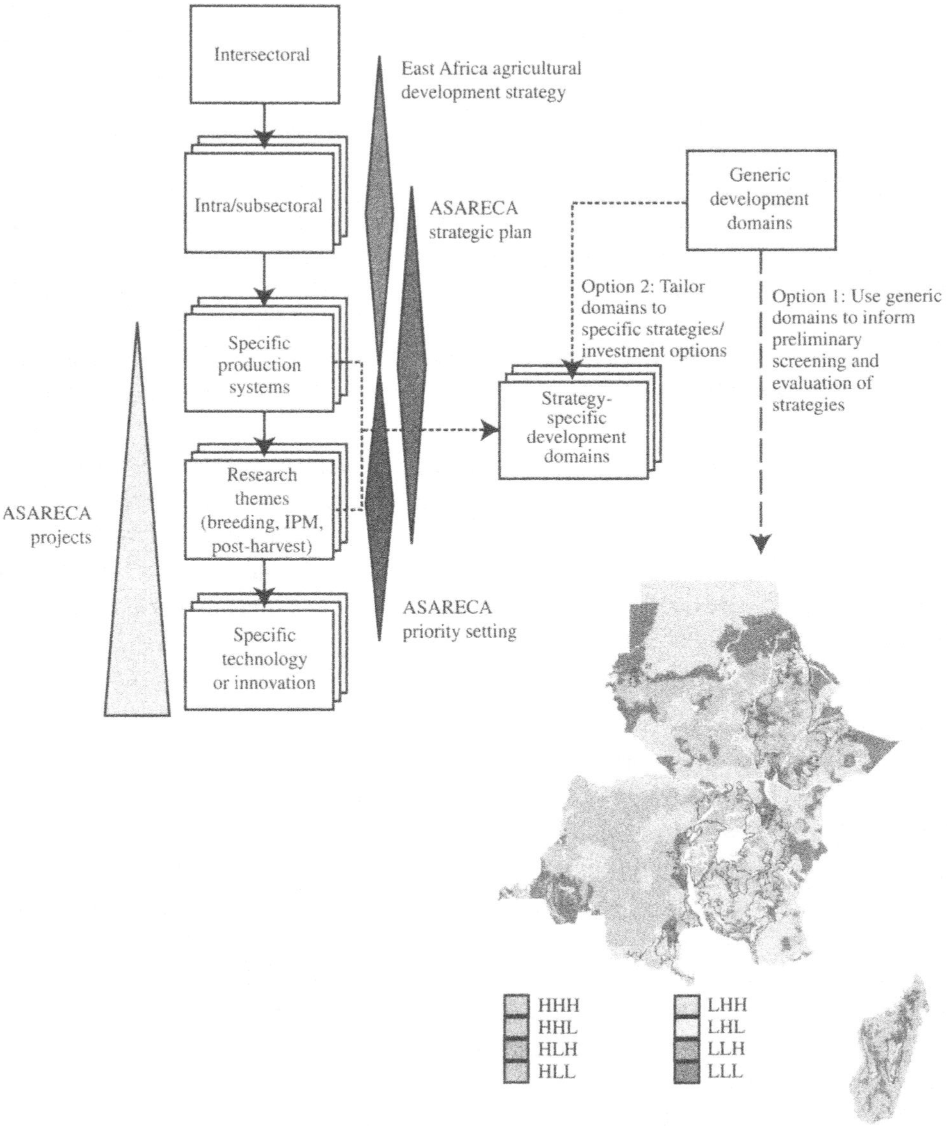

(continued)

Box A.1—Continued

is to get a clearer spatial perspective of the general subsectoral development options; for example, what might be the spatial implications of pursuing extensive livestock, versus high-input cereal, versus staple food crop expansion strategies?

As specific strategies are elaborated and refined, corresponding degrees of specificity can be introduced into the domain definitions. For example, for a maize intensification strategy, we can generate a more targeted agricultural potential map highlighting the conditions of temperature, rainfall, slope, soil and drainage properties, and even pest and weed incidence that best match the specific needs of maize intensification. Similarly, if the intent was to promote maize intensification to expand regional trade, a market access map could be generated highlighting those areas with better access to regional trade routes and local border crossings. Furthermore, having identified a better-targeted subset of areas across ECA where an individual strategy or group of strategies may be most promising, it is more feasible—and likely necessary—to define domains with higher levels of spatial resolution, potentially including other higher-resolution datasets that may be available at more local scales.

The analytical spans of the two additional strategic scales of inquiry being targeted by this study, the ASARECA Research for Development Strategy and the ASARECA Priority Setting, are illustrated in the figure by the middle and lowest diamonds, respectively. The additional levels of information required to address each subsequent scale enhance the relevance of further elaborating the generic domains to arrive at (perhaps multiple) domain schemes. Such domain schemes better reflect the greater diversity and specificity of those production environments and technology combinations requiring scrutiny.

Although the generic domains offer useful and regionally consistent spatial perspectives on targeting and investment strategies for ASARECA networks, programs, and projects, managers of individual programs might want to quickly add nested or hierarchical domain schemes that capture the production and marketing constraints and opportunities that best match their own situations. To foster this approach, all generic domain spatial and tabular data will be distributed freely.

pests, diseases, floods, droughts, erosion hazards, and so on. Within ECA, dominated by a smallholder, subsistence-oriented agricultural production base, some of the most binding constraints to improved agricultural production potential have been recognized as the amount and variability of water supply, the biotic pressure of pest and diseases, and soil fertility.

In reviewing which of these elements were critical for inclusion in the ECA agricultural potential measure, the expert consultation came to the following conclusions:

- Annual rainfall total alone is not sufficient to define agricultural potential in ECA, because (1) the seasonality (for example, there are both unimodal and bimodal rainfall patterns in the region) and shorter-term variability of rainfall within a year have important implications on potential production options and productivity and (2) there are significant areas in the region where the agricultural production does not depend directly on rainfall, but indirectly through irrigation, access to surface-water bodies, seasonally high water tables, and groundwater resources. One example given was cultivation in wet soils as lake levels recede during the dry season.

- Agricultural potential must include some recognition of the variability in soil quality, especially soil fertility, across the region. Organic matter, pH, soil depth, and texture were considered the most critical indicators of soil quality.
- There is already evidence of changing patterns of rainfall variability, including increased severity of extreme events, as well as longer-term temperature and rainfall trends, associated with climate change. Because the domains are helping to assess the potential for future agricultural development, where these changes are likely to be significant they should be taken into account.
- From a planning perspective, there are some types of area that should not be promoted for agriculture-based livelihood options (although other types of livelihood may be feasible). In particular it was recommended that protected areas be omitted from agricultural development domains. Steeply sloping and high-elevation areas were also suggested for omission.
- Areas of endemic human and animal disease—tsetse being a very specific and important example in ECA—were identified as being key biophysical determinants of agricultural potential that should be taken into account.

In response to these recommendations and balanced with the availability of consistent and reliable data at a regional scale, a three-level scheme was developed for assessing agricultural potential in ECA, in terms of three classes: high (medium to high) potential, low (low to medium) potential, and not feasible.[29] The scheme is shown diagrammatically in Figure A.1. There are three elements of the classification process. At the highest level, some areas are excluded from further consideration on the grounds of landscape or land use criteria. At the next level, rules are applied that reflect various water availability possibilities, and at the final level, rules to account for soil quality are applied. The above sets of rules were applied in turn to each 20 × 20-kilometer pixel in the ECA region.

Landscape/Land Use Rules. A pixel was considered not suitable for agricultural development if any of the following was true: elevation is greater than 2,800 meters, population density is less than two persons per square kilometer, or the pixel is within a protected area. The population rule was introduced to capture areas that might later prove to be feasible because of water and soil conditions, but where there are obviously significant constraints to human settlement, such as tsetse infestation.

Water Availability Rules. Pixels that contained more than 15 percent of the area irrigated, that were within a specified distance of a surface water body, or river or that had a length of growing period (LGP) of five months or longer were classified as being of high water availability. Pixels of between two and five months' LGP were classified as being of low agricultural potential, and those of less than two months' LGP were classified as not suitable.[30]

[29]Three classes may seem excessively aggregated, but we are attempting to minimize the final number of development domain categories, of which agricultural potential is just a single layer. The final number of classes in the development domain surface is the product of the classes in each component layer (although not all combinations may actually occur).

[30]LGP, defined as the number of days in a year when sufficient water is available in the soil profile to support plant growth, is a shorthand way of capturing the complex interaction of rainfall, potential evapotranspiration, and soil properties. LGP has recently been calculated and mapped globally at a scale of 30 minutes (Fischer et al. 2001). The LGP calculation was made based on 30 years of spatially interpolated monthly rainfall data (New, Hulme, and Jones 2000) and measures of both average number of days per year and the standard error

Figure A.1 ECA development domains: Generation of generalized agricultural potential layer

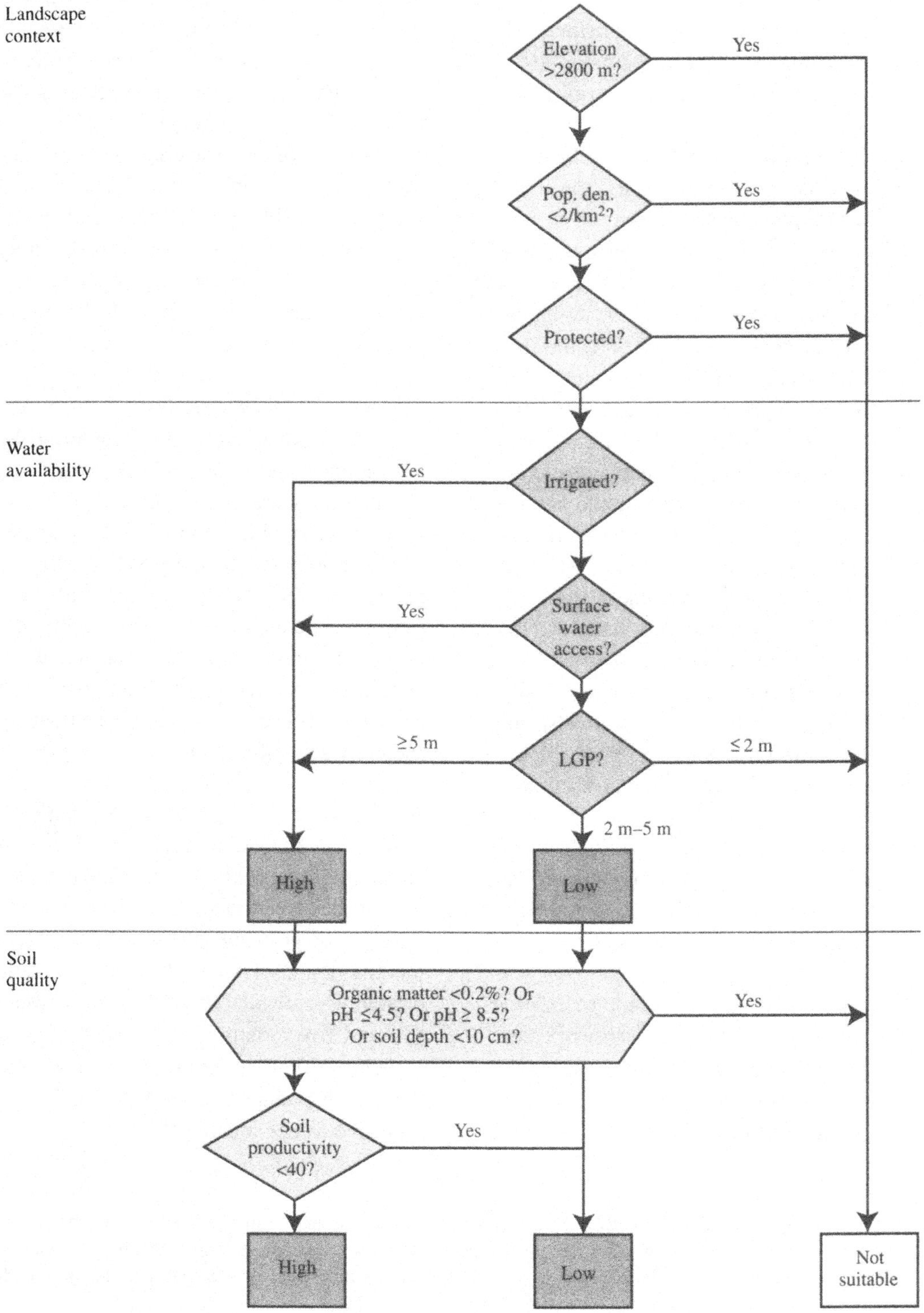

Soil Quality. Pixels for which the topsoil was reported as having an organic matter content less than 0.2 percent, a pH less than 4.5 or greater than 8.5, or a soil depth of less than 10 centimeters were classified as not feasible. Thereafter, pixels classified as high potential from a water availability perspective, but whose soil productivity index was less than 40, were assigned low potential.

Potential Market Access

As with agricultural potential, this is a complex factor, definable in a variety of ways. Opportunities for gathering market information, obtaining credit, buying inputs, selling outputs, and so on depend on a wide range of socioeconomic, institutional, and cultural factors that are not necessarily associated with settlement size or the connectivity among locations. However, we are usually forced to adopt a practical metric of market access, such as physical distance or (with greater analytical effort) travel time between production locations and predefined market locations. Such market locations are usually determined using a settlement population size criteria (for example, more than 100,000) or by assuming, say, that all district administrative centers serve as market towns. Such simple rules ignore the very real barriers to market access imposed by missing or ineffective markets as a consequence of poor information, infrastructure, and marketing or market regulating institutions. This method may also miss the importance of other spatial patterns of marketing chains (for example, cotton to ginneries rather than trading centers). Furthermore, market opportunities often have seasonal patterns that are difficult to quantify and represent in spatially explicit ways. Nonetheless, geographical locations can be identified for different market opportunities and physical access does capture some important aspects of the potential for market engagement, especially in rural areas.

In the technical consultations, much of the discussion on market access focused on the conceptual and methodological challenges of deriving market accessibility metrics (for example, the options for representing and accounting for travel to market from farming locations not directly served by transport infrastructure). One recurring issue was the necessity to specify what types of markets are actually being served. The notion of deriving separate market access surfaces for different types of market—local, capital city, export corridors, trans-shipment points—was also proposed. This approach was very congruent with the goal of generating development domains that are specific to certain strategies; for example, an export corridor and border crossing point may be most relevant for strategies involving tradable commodities.

Here we conceptually decompose the range of market opportunities into major subgroups, each with a spatially definable set of market targets (that is, physical loci of marketing opportunities). These market subgroups are local trade, subregional trading centers, central urban markets, regional overland trade flows, and international "fresh markets" accessed by airports. Based on data on road location and quality, terrain slope, and off-road landcover, we model the travel times to these locations from all other points in the region. We then characterize

of LGP were generated. A growing period was defined to begin when rainfall exceeds half the potential rate of evapotranspiration, and to finish when a prespecified amount of soil water left in storage after the end of the rainfall period had evaporated. That amount depended on average soil depth and texture within each 30-minute grid cell. When multiple growing seasons occur within a year (for example, in the northern crescent around Lake Victoria), the sum of both growing periods was used to assess total growing period (however, the LGP database records the start and end date and length of each individual growing period to support more detailed analysis).

Figure A.2 ECA development domains: Generation of composite potential market access layer

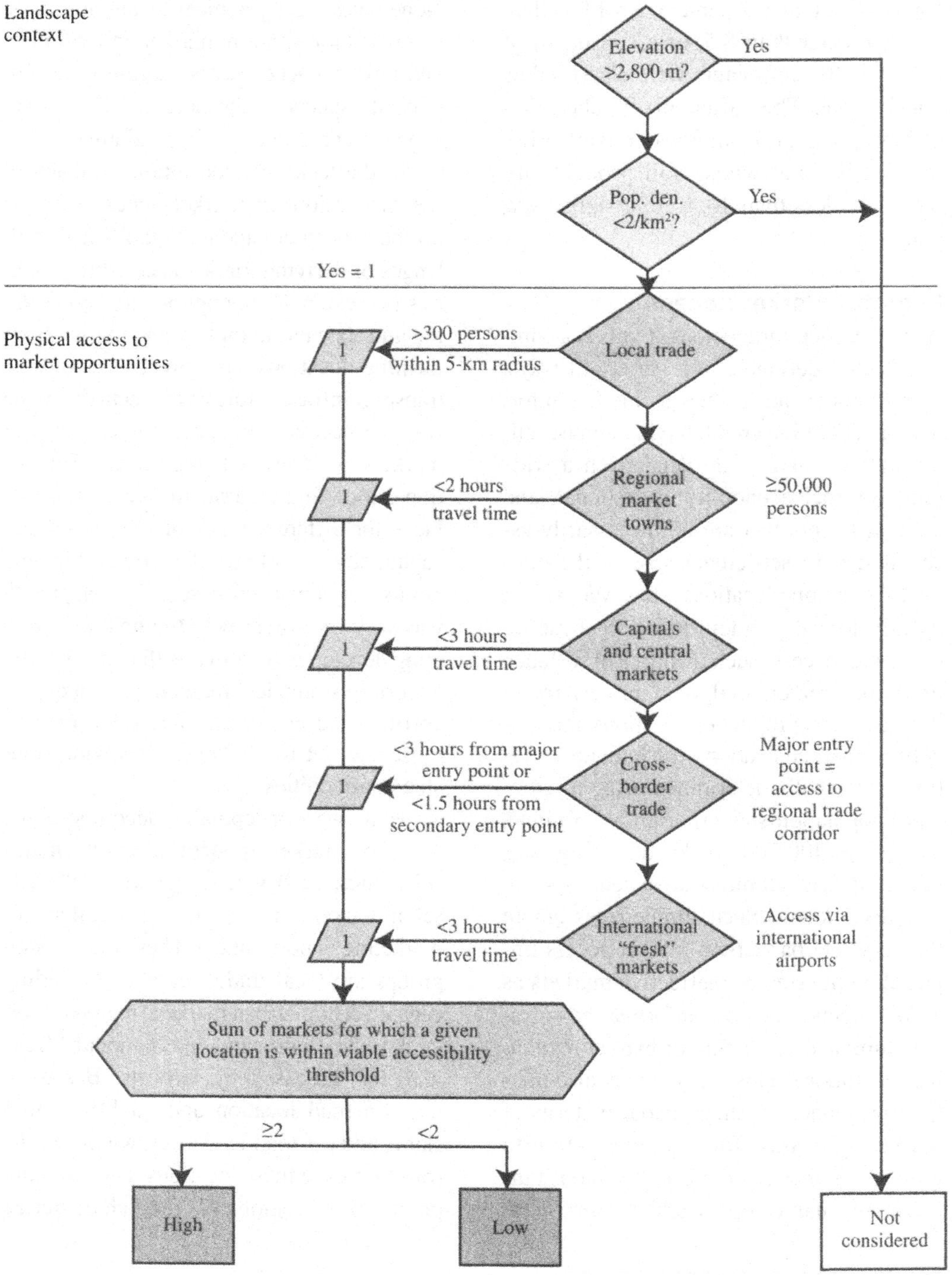

the relative market access for a given locale on the basis of the set of markets in which it can feasibly participate.

Figure A.2 illustrates the sets and rules applied to determine types and levels of market access. We first removed from consideration those areas where trade can safely be presumed not to exist: noninhabitable and uninhabited places. Remaining areas are then considered in terms of whether they fall within a given threshold of accessibility (measured in travel time) to each set of geographical market targets.

The opportunity for local trade is considered high for those areas with 300 or more persons within a 5-kilometer radius (calculations are based on the population density grids described in next section). Another set of opportunities is provided by local trading centers. These are identified as settlements of 50,000 or more people; access is considered high for those areas within two hours of travel time. Capitals and other major urban centers are considered as another set of targets. Areas within three hours of travel time are considered to have high access to these markets.

The mapping of cross-border trade flows is challenging. We drew on expert consensus to identify entry points to major and secondary overland trade corridors in the region. Areas within 3 hours of a major portal (for example, the Tororo border crossing on the Kampala-Nairobi-Mombassa trade corridor) were considered to have high access; areas within 1.5 hours of a secondary access point (for example, near Arua on the Uganda–Democratic Republic of Congo [DRC] border) were likewise considered to have high access.

Finally, access to high-value, fresh produce market opportunities in Europe and elsewhere is presumed to occur primarily via international air travel. As such, areas within three hours of a major international airport were classified as high access. Figure A.3 illustrates the five resulting market access surfaces and the composite surface that indicates the total number of markets to which any location in ECA has access.

If an area fell within the viable thresholds of two or more markets, it was classified as medium-high potential; otherwise, medium-low. The resulting patterns of classified areas conform to the broadly held understanding of regional market opportunities: areas proximate to the major trade corridors show up as high access, as do areas surrounding capitals in the high-density highlands of Ethiopia, Rwanda, and Burundi. Elsewhere high-access areas are more restricted to those areas contiguous with urban centers.

Population Density

The land to labor ratio has been theorized to have consequences for land management and other production technology strategies (following Boserup's [1965] theory of induced innovation). That is to say, holding other factors constant, farmers in areas of high population density are more likely to adopt labor-intensive production strategies than are those in areas of low density. As such, population density is a useful organizing frame for examining land management decisions.

Because domains will be used to help gauge the potential effects of innovations in rural production and natural resource management strategies, it makes sense to examine the patterns of rural population densities only, to the extent that they can be isolated spatially from urban areas. A global dataset of gridded population, segregated into urban and rural components is available under the Global Rural-Urban Mapping Project (CIESIN et al. 2004). This dataset uses a variety of sources to identify the spatial and quantitative magnitude of urban populations: sizes of urban areas are derived primarily from satellite imagery and moderate-scale navigational charts; population sizes for these urban areas are collected from national reporting agencies. Where no urban boundaries are available but urban populations are reported, a spatial unit is defined on the basis of settlement size-area relationships defined statistically by region. Likewise, where no population figure is available for a recognized urban area, a regionally appropriate estimate is assigned. Urban population is thus netted out of total population reported for each administrative unit (for example, state, region, or district). The population assigned to each pixel within the unit is the result of assuming a uniform density of urban population in urban areas, and

Figure A.3 Five potential market access layers and composite of number of accessible markets

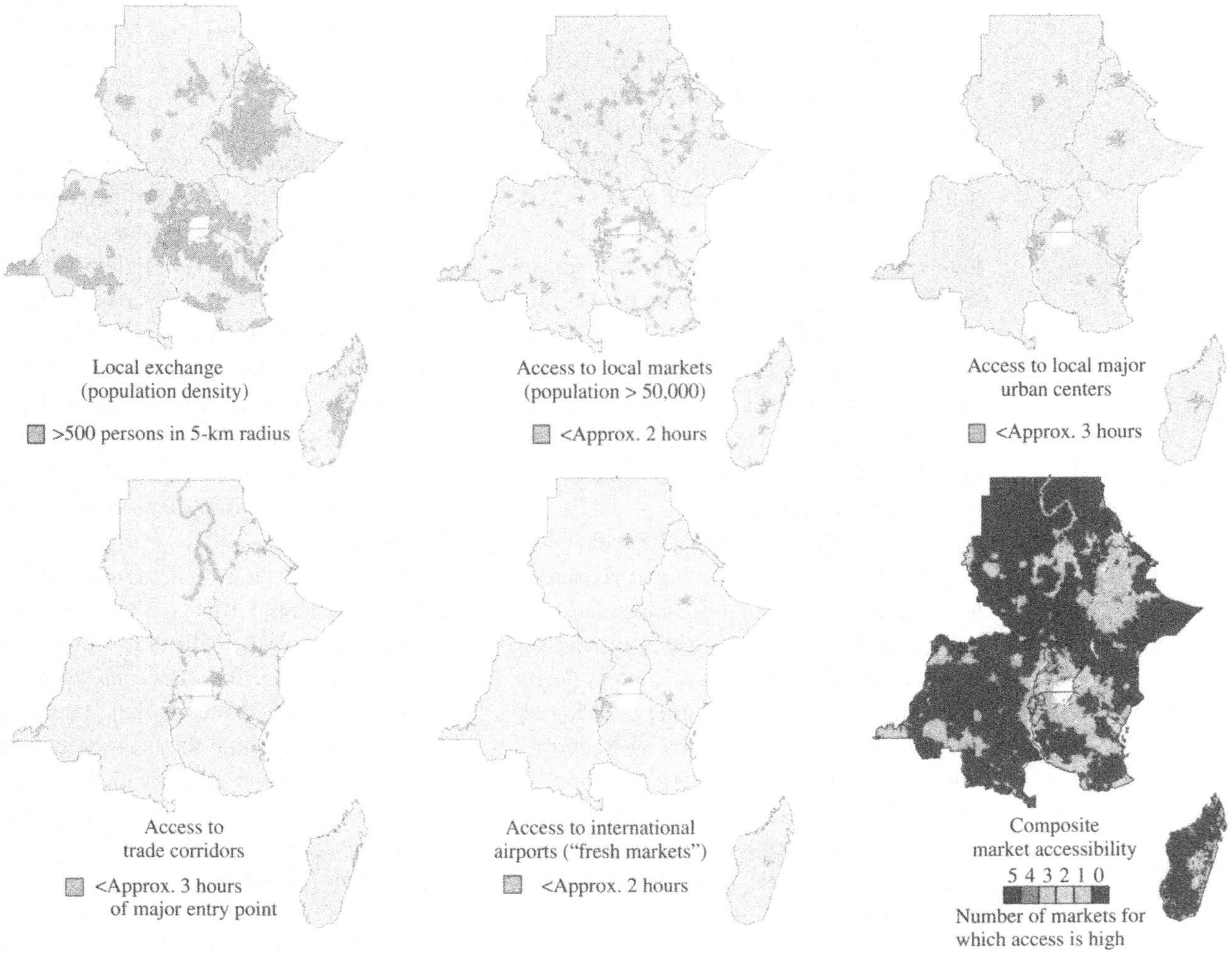

rural populations across the remaining, non-urban areas within the unit. Because reporting units are increasingly disaggregated (mapped reporting units in Uganda, for example, are at the parish scale), this "semi-modeled" population distribution is of adequate resolution to indicate meaningful subregional patterns of population density.

As far as segregating areas of high and low densities, research is still needed to validate density thresholds for these categories. Nonetheless, 100 persons per square kilometer is a value frequently used in development studies to identify high-density areas. Furthermore, mapped delineations of these areas tend to correspond well with expert perceptions of where densely populated areas are located. Given this situation, we adhere to the definition of 100 or more persons per square kilometer for high-density areas.

The temperate and subtropical highland areas of Burundi, Ethiopia, Kenya, Rwanda, and Uganda are the predominant high-density areas of the region. Smaller pockets of high population densities are found in northeastern DRC and areas bordering Lake Victoria.

Distributions of Land, Cropland, and Population by Development Domain from a Country Perspective

Figure 3.5 is the map of the development domains presented in the main report. This map is the result of intersecting the three separate maps of agricultural potential, market access, and population density, each classified into high and low categories. In the main report, we present and briefly discuss those domains from a regional perspective. Here we show the domain attributes by country (Tables A.1a–d).

Table A.1a Land area distribution across development domains by country and region (percent)

Domain	Burundi	Eritrea	Ethiopia	Kenya	Madagascar	Rwanda	Sudan	Uganda	DRC	Tanzania	ECA
HHH	34	1	4	4	1	24	0	16	1	1	2
HHL	2	3	1	2	2	5	2	7	2	5	2
HLH	25	0	9	4	1	11	0	13	1	3	3
HLL	10	20	32	17	62	6	22	39	61	32	38
LHH	19	2	1	3	1	37	0	3	0	0	1
LHL	0	8	0	2	1	1	2	1	1	3	1
LLH	5	0	4	2	1	7	0	1	0	1	1
LLL	0	57	30	33	27	0	24	7	11	11	21
Not included	4	9	18	33	5	10	49	13	24	44	31
Total	100	100	100	100	100	100	100	100	100	100	100

Notes: The three factors making up the domain definitions are agricultural potential, market access, and population density. H = high; L = low.

Table A.1b Rural population distribution across development domains by country and region (percent)

Domain	Burundi	Eritrea	Ethiopia	Kenya	Madagascar	Rwanda	Sudan	Uganda	DRC	Tanzania	ECA
HHH	43	7	13	22	5	26	2	40	5	7	14
HHL	1	4	2	2	6	1	8	5	5	8	4
HLH	20	0	30	20	7	8	0	22	8	12	16
HLL	5	13	19	6	48	2	38	15	55	38	28
LHH	22	21	3	18	5	50	4	10	2	1	7
LHL	0	15	1	3	4	0	9	1	1	4	3
LLH	5	0	11	12	4	7	1	2	3	3	6
LLL	0	37	14	12	19	0	34	2	15	15	15
Not included	3	3	8	4	2	6	4	2	5	10	6
Total	100	100	100	100	100	100	100	100	100	100	100

Notes: The three factors making up the domain definitions are agricultural potential, market access, and population density. H = high; L = low.

Table A.1c Cropland distribution across development domains by country and region (percent)

Domain	Burundi	Eritrea	Ethiopia	Kenya	Madagascar	Rwanda	Sudan	Uganda	DRC	Tanzania	ECA
HHH	39	0	11	10	1	23	1	24	2	2	7
HHL	3	2	3	2	4	3	5	7	1	6	4
HLH	24	0	24	9	2	9	0	13	4	3	9
HLL	7	62	27	14	64	2	59	38	53	32	39
LHH	19	2	2	8	1	46	1	5	1	0	3
LHL	1	8	1	3	2	1	3	1	1	4	2
LLH	5	0	8	6	2	6	0	1	3	1	3
LLL	0	24	13	34	21	0	12	3	18	15	16
Not included	3	2	11	13	4	10	19	8	18	36	17
Total	100	100	100	100	100	100	100	100	100	100	100

Notes: The three factors making up the domain definitions are agricultural potential, market access, and population density. H = high; L = low.

Table A.1d Pasture distribution across development domains by country and region (percent)

Domain	Burundi	Eritrea	Ethiopia	Kenya	Madagascar	Rwanda	Sudan	Uganda	DRC	Tanzania	ECA
HHH	38	1	4	4	1	28	0	20	2	2	3
HHL	3	4	2	2	2	5	4	7	2	6	3
HLH	26	0	9	4	1	8	0	13	3	3	3
HLL	8	24	30	18	60	5	43	37	52	34	39
LHH	16	3	1	2	1	41	1	4	1	0	1
LHL	0	13	0	1	1	1	3	1	1	3	2
LLH	5	0	4	2	1	6	0	1	1	1	1
LLL	0	48	31	32	30	0	25	5	18	11	25
Not included	4	6	19	36	4	8	24	10	21	40	22
Total	100	100	100	100	100	100	100	100	100	100	100

Notes: The three factors making up the domain definitions are agricultural potential, market access, and population density. H = high; L = low.

APPENDIX B

Economywide Multimarket Model

Most multimarket models are partial equilibrium ones focusing on certain sectors in the economy. Although the economywide multimarket model (EMM) developed for this project focuses on agriculture, the rest of economic activities are included as two aggregated sectors. Thus the current model can partially capture general equilibrium linkages of the economy. The model includes 10 countries, and subnational disaggregation can be done with data availability. There are 31 disaggregated agricultural commodities, 25 crop commodities, 6 livestock commodities, and 2 aggregate nonagricultural commodities.

The following commodities are included in the model: maize, rice, wheat, sorghum, barley, millet, oats, other cereals, potatoes, sweet potatoes, cassava, other root crops, beans, peas, groundnuts, sesame seed, other oil crops, vegetable oil, sugar, vegetables, bananas, fruits, coffee, tea, spices and beverages, cotton, bovine meat, goat and mutton meat, poultry and eggs, other meat, milk, and fish.

Supply Functions

Consistent with most multimarket model setups, the supply function, instead of production function, is used to capture each representative producer's response to market. In the crop subsectors, the subnational supply functions for the 25 crop products have two components: (1) yield functions that are used to capture supply response to own prices, given farm area allocated to this crop and (2) land-allocation functions that are functions of all prices and hence are responsive to changing profitability across different crops, given the total available land. The supply functions for the livestock and nonagricultural sectors are dependent on the prices for all commodities and productivity parameters.

If data were available, differences in technology (for example, modern input uses) and uses of inputs in the production would be modeled either as different yield functions for a given crop or as endogenous variables in the yield function. For example, if there were data about fertilizer use and irrigation by area and crop, it would be possible to have four types of yield functions reflecting the combinations of the inputs (fertilizer use only, irrigation only, fertilizer and irrigation combined, and without using any modern input) or to have fertilizer prices included in the yield functions. In the current version of EMM, both differences in technology and uses of inputs (except for land) are not explicitly included because of data constraints. For the total land constraint in a specific country, the model imposes certain constraints on the elasticities employed in the supply functions to avoid a simultaneous increase in the output of all products within a country in a given year. Table B.1 shows supply elasticities, Table B.2 shows growth rates in commodity subsectors, and Table B.3 lists levels of production of agricultural commodities in ECA.

Table B.1 Commodity supply price elasticities

	Maize	Rice	Wheat	Sorghum	Barley	Other cereals	Millet	Oats	Potatoes	Sweet potatoes	Cassava
Burundi											
Maize	0.089	−0.053	0.000	−0.001			0.000		0.000	−0.013	−0.008
Rice	−0.028	0.197	0.000	−0.001			0.000		0.000	−0.013	−0.008
Wheat	−0.028	−0.053	0.147	−0.001			0.000		0.000	−0.013	−0.008
Sorghum	−0.028	−0.053	0.000	0.146			0.000		0.000	−0.013	−0.008
Millet	−0.028	−0.053	0.000	−0.001			0.148		0.000	−0.013	−0.008
Potatoes	−0.028	−0.053	0.000	−0.001			0.000		0.148	−0.013	−0.008
Sweet potatoes	−0.028	−0.053	0.000	−0.001			0.000		0.000	0.157	−0.008
Cassava	−0.028	−0.053	0.000	−0.001			0.000		0.000	−0.013	0.153
Other roots	−0.028	−0.053	0.000	−0.001			0.000		0.000	−0.013	−0.008
Beans	−0.028	−0.053	0.000	−0.001			0.000		0.000	−0.013	−0.008
Peas	−0.028	−0.053	0.000	−0.001			0.000		0.000	−0.013	−0.008
Groundnuts	−0.057	−0.106	0.000	−0.002			0.000		−0.001	−0.026	−0.017
Other oil crops	−0.057	−0.106	0.000	−0.002			0.000		−0.001	−0.026	−0.017
Vegetables	−0.028	−0.053	0.000	−0.001			0.000		0.000	−0.013	−0.008
Bananas	−0.028	−0.053	0.000	−0.001			0.000		0.000	−0.013	−0.008
Fruits	−0.028	−0.053	0.000	−0.001			0.000		0.000	−0.013	−0.008
Coffee	−0.028	−0.053	0.000	−0.001			0.000		0.000	−0.013	−0.008
Tea	−0.028	−0.053	0.000	−0.001			0.000		0.000	−0.013	−0.008
Cotton lint	−0.028	−0.053	0.000	−0.001			0.000		0.000	−0.013	−0.008
DRC											
Maize	0.094	−0.002		0.000			−0.001		0.000	0.000	−0.122
Rice	−0.011	0.144		0.000			−0.001		0.000	0.000	−0.122
Sorghum	−0.011	−0.002		0.148			−0.001		0.000	0.000	−0.122
Millet	−0.011	−0.002		0.000			0.140		0.000	0.000	−0.122
Potatoes	−0.011	−0.002		0.000			−0.001		0.148	0.000	−0.122
Sweet potatoes	−0.011	−0.002		0.000			−0.001		0.000	0.148	−0.122
Cassava	−0.011	−0.002		0.000			−0.001		0.000	0.000	0.183
Other roots	−0.011	−0.002		0.000			−0.001		0.000	0.000	−0.122
Peas	−0.011	−0.002		0.000			−0.001		0.000	0.000	−0.122
Groundnuts	−0.023	−0.005		0.000			−0.002		0.000	0.000	−0.244
Raw sugar	−0.011	−0.002		0.000			−0.001		0.000	0.000	−0.122
Vegetables	−0.011	−0.002		0.000			−0.001		0.000	0.000	−0.122
Bananas	−0.011	−0.002		0.000			−0.001		0.000	0.000	−0.122
Fruits	−0.011	−0.002		0.000			−0.001		0.000	0.000	−0.122
Coffee	−0.011	−0.002		0.000			−0.001		0.000	0.000	−0.122
Tea	−0.011	−0.002		0.000			−0.001		0.000	0.000	−0.122
Spices and beverages	−0.011	−0.002		0.000			−0.001		0.000	0.000	−0.122

Other roots	Beans	Peas	Groundnuts	Sesame seeds	Other oil crops	Raw sugar	Vegetables	Bananas	Fruits	Coffee	Tea	Spices and beverages	Cotton lint
0.000	−0.001	0.000	−0.002		−0.001		−0.015	−0.023	−0.005	−0.004	−0.001		−0.001
0.000	−0.001	0.000	−0.002		−0.001		−0.015	−0.023	−0.005	−0.004	−0.001		−0.001
0.000	−0.001	0.000	−0.002		−0.001		−0.015	−0.023	−0.005	−0.004	−0.001		−0.001
0.000	−0.001	0.000	−0.002		−0.001		−0.015	−0.023	−0.005	−0.004	−0.001		−0.001
0.000	−0.001	0.000	−0.002		−0.001		−0.015	−0.023	−0.005	−0.004	−0.001		−0.001
0.000	−0.001	0.000	−0.002		−0.001		−0.015	−0.023	−0.005	−0.004	−0.001		−0.001
0.000	−0.001	0.000	−0.002		−0.001		−0.015	−0.023	−0.005	−0.004	−0.001		−0.001
0.000	−0.001	0.000	−0.002		−0.001		−0.015	−0.023	−0.005	−0.004	−0.001		−0.001
0.147	−0.001	0.000	−0.002		−0.001		−0.015	−0.023	−0.005	−0.004	−0.001		−0.001
0.000	0.140	0.000	−0.002		−0.001		−0.015	−0.023	−0.005	−0.004	−0.001		−0.001
0.000	−0.001	0.146	−0.002		−0.001		−0.015	−0.023	−0.005	−0.004	−0.001		−0.001
−0.001	−0.002	0.000	0.293		−0.003		−0.030	−0.046	−0.009	−0.008	−0.003		−0.001
−0.001	−0.002	0.000	−0.004		0.274		−0.030	−0.046	−0.009	−0.008	−0.003		−0.001
0.000	−0.001	0.000	−0.002		−0.001		0.156	−0.023	−0.005	−0.004	−0.001		−0.001
0.000	−0.001	0.000	−0.002		−0.001		−0.015	0.153	−0.005	−0.004	−0.001		−0.001
0.000	−0.001	0.000	−0.002		−0.001		−0.015	−0.023	0.154	−0.004	−0.001		−0.001
0.000	−0.001	0.000	−0.002		−0.001		−0.015	−0.023	−0.005	0.151	−0.001		−0.001
0.000	−0.001	0.000	−0.002		−0.001		−0.015	−0.023	−0.005	−0.004	0.149		−0.001
0.000	−0.001	0.000	−0.002		−0.001		−0.015	−0.023	−0.005	−0.004	−0.001		0.147
0.000		0.000	−0.024			0.000	−0.002	−0.003	−0.001	0.000	0.000	−0.021	
0.000		0.000	−0.024			0.000	−0.002	−0.003	−0.001	0.000	0.000	−0.021	
0.000		0.000	−0.024			0.000	−0.002	−0.003	−0.001	0.000	0.000	−0.021	
0.000		0.000	−0.024			0.000	−0.002	−0.003	−0.001	0.000	0.000	−0.021	
0.000		0.000	−0.024			0.000	−0.002	−0.003	−0.001	0.000	0.000	−0.021	
0.000		0.000	−0.024			0.000	−0.002	−0.003	−0.001	0.000	0.000	−0.021	
0.000		0.000	−0.024			0.000	−0.002	−0.003	−0.001	0.000	0.000	−0.021	
0.148		0.000	−0.024			0.000	−0.002	−0.003	−0.001	0.000	0.000	−0.021	
0.000		0.147	−0.024			0.000	−0.002	−0.003	−0.001	0.000	0.000	−0.021	
0.000		0.000	0.248			0.000	−0.004	−0.006	−0.003	0.000	0.000	−0.042	
0.000		0.000	−0.024			0.147	−0.002	−0.003	−0.001	0.000	0.000	−0.021	
0.000		0.000	−0.024			0.000	0.149	−0.003	−0.001	0.000	0.000	−0.021	
0.000		0.000	−0.024			0.000	−0.002	0.147	−0.001	0.000	0.000	−0.021	
0.000		0.000	−0.024			0.000	−0.002	−0.003	0.151	0.000	0.000	−0.021	
0.000		0.000	−0.024			0.000	−0.002	−0.003	−0.001	0.148	0.000	−0.021	
0.000		0.000	−0.024			0.000	−0.002	−0.003	−0.001	0.000	0.148	−0.021	
0.000		0.000	−0.024			0.000	−0.002	−0.003	−0.001	0.000	0.000	0.196	

(continued)

Table B.1—Continued

	Maize	Rice	Wheat	Sorghum	Barley	Other cereals	Millet	Oats	Potatoes	Sweet potatoes	Cassava
Eritrea											
Maize	0.140		−0.022	−0.002	−0.019	−0.015					
Wheat	−0.003		0.150	−0.002	−0.019	−0.015					
Sorghum	−0.003		−0.022	0.142	−0.019	−0.015					
Barley	−0.003		−0.022	−0.002	0.152	−0.015					
Other cereals	−0.003		−0.022	−0.002	−0.019	0.140					
Beans	−0.003		−0.022	−0.002	−0.019	−0.015					
Peas	−0.003		−0.022	−0.002	−0.019	−0.015					
Sesame seeds	−0.007		−0.045	−0.003	−0.038	−0.030					
Other oil crops	−0.007		−0.045	−0.003	−0.038	−0.030					
Vegetables	−0.003		−0.022	−0.002	−0.019	−0.015					
Fruits	−0.003		−0.022	−0.002	−0.019	−0.015					
Spices and beverages	−0.003		−0.022	−0.002	−0.019	−0.015					
Ethiopia											
Maize	0.020		0.000	−0.001	0.000	0.000					
Wheat	−0.004		0.147	−0.001	0.000	0.000					
Sorghum	−0.004		0.000	0.114	0.000	0.000					
Barley	−0.004		0.000	−0.001	0.148	0.000					
Other cereals	−0.004		0.000	−0.001	0.000	0.119					
Peas	−0.004		0.000	−0.001	0.000	0.000					
Vegetables	−0.004		0.000	−0.001	0.000	0.000					
Spices and beverages	−0.004		0.000	−0.001	0.000	0.000					
Cotton lint	−0.004		0.000	−0.001	0.000	0.000					
Kenya											
Maize	0.173	−0.003	0.000	0.000			0.000		−0.001	0.000	−0.003
Rice	−0.005	0.147	0.000	0.000			0.000		−0.001	0.000	−0.003
Wheat	−0.005	−0.003	0.149	0.000			0.000		−0.001	0.000	−0.003
Sorghum	−0.005	−0.003	0.000	0.154			0.000		−0.001	0.000	−0.003
Millet	−0.005	−0.003	0.000	0.000			0.148		−0.001	0.000	−0.003
Potatoes	−0.005	−0.003	0.000	0.000			0.000		0.148	0.000	−0.003
Sweet potatoes	−0.005	−0.003	0.000	0.000			0.000		−0.001	0.148	−0.003
Cassava	−0.005	−0.003	0.000	0.000			0.000		−0.001	0.000	0.147
Peas	−0.005	−0.003	0.000	0.000			0.000		−0.001	0.000	−0.003
Groundnuts	−0.010	−0.006	−0.001	0.000			0.000		−0.001	−0.001	−0.005
Sesame seeds	−0.010	−0.006	−0.001	0.000			0.000		−0.001	−0.001	−0.005
Other oil crops	−0.010	−0.006	−0.001	0.000			0.000		−0.001	−0.001	−0.005
Raw sugar	−0.005	−0.003	0.000	0.000			0.000		−0.001	0.000	−0.003
Vegetables	−0.005	−0.003	0.000	0.000			0.000		−0.001	0.000	−0.003
Bananas	−0.005	−0.003	0.000	0.000			0.000		−0.001	0.000	−0.003
Fruits	−0.005	−0.003	0.000	0.000			0.000		−0.001	0.000	−0.003
Coffee	−0.005	−0.003	0.000	0.000			0.000		−0.001	0.000	−0.003
Tea	−0.005	−0.003	0.000	0.000			0.000		−0.001	0.000	−0.003
Spices and beverages	−0.005	−0.003	0.000	0.000			0.000		−0.001	0.000	−0.003

Other roots	Beans	Peas	Groundnuts	Sesame seeds	Other oil crops	Raw sugar	Vegetables	Bananas	Fruits	Coffee	Tea	Spices and beverages	Cotton lint
	−0.001	−0.013		−0.001	−0.006		−0.004		−0.002			−0.035	
	−0.001	−0.013		−0.001	−0.006		−0.004		−0.002			−0.035	
	−0.001	−0.013		−0.001	−0.006		−0.004		−0.002			−0.035	
	−0.001	−0.013		−0.001	−0.006		−0.004		−0.002			−0.035	
	−0.001	−0.013		−0.001	−0.006		−0.004		−0.002			−0.035	
	0.147	−0.013		−0.001	−0.006		−0.004		−0.002			−0.035	
	−0.001	0.109		−0.001	−0.006		−0.004		−0.002			−0.035	
	−0.001	−0.027		0.284	−0.012		−0.008		−0.004			−0.071	
	−0.001	−0.027		−0.002	0.206		−0.008		−0.004			−0.071	
	−0.001	−0.013		−0.001	−0.006		0.149		−0.002			−0.035	
	−0.001	−0.013		−0.001	−0.006		−0.004		0.153			−0.035	
	−0.001	−0.013		−0.001	−0.006		−0.004		−0.002			0.232	
		0.000					−0.001					−0.001	0.000
		0.000					−0.001					−0.001	0.000
		0.000					−0.001					−0.001	0.000
		0.000					−0.001					−0.001	0.000
		0.000					−0.001					−0.001	0.000
		0.144					−0.001					−0.001	0.000
		0.000					0.125					−0.001	0.000
		0.000					−0.001					0.149	0.000
		0.000					−0.001					−0.001	0.143
		0.000	0.000	0.000	0.000	−0.003	−0.002	−0.002	−0.046	−0.020	−0.100	−0.002	
		0.000	0.000	0.000	0.000	−0.003	−0.002	−0.002	−0.046	−0.020	−0.100	−0.002	
		0.000	0.000	0.000	0.000	−0.003	−0.002	−0.002	−0.046	−0.020	−0.100	−0.002	
		0.000	0.000	0.000	0.000	−0.003	−0.002	−0.002	−0.046	−0.020	−0.100	−0.002	
		0.000	0.000	0.000	0.000	−0.003	−0.002	−0.002	−0.046	−0.020	−0.100	−0.002	
		0.000	0.000	0.000	0.000	−0.003	−0.002	−0.002	−0.046	−0.020	−0.100	−0.002	
		0.000	0.000	0.000	0.000	−0.003	−0.002	−0.002	−0.046	−0.020	−0.100	−0.002	
		0.000	0.000	0.000	0.000	−0.003	−0.002	−0.002	−0.046	−0.020	−0.100	−0.002	
		0.151	0.000	0.000	0.000	−0.003	−0.002	−0.002	−0.046	−0.020	−0.100	−0.002	
		−0.001	0.296	0.000	−0.001	−0.005	−0.004	−0.004	−0.092	−0.040	−0.199	−0.004	
		−0.001	0.000	0.297	−0.001	−0.005	−0.004	−0.004	−0.092	−0.040	−0.199	−0.004	
		−0.001	0.000	0.000	0.297	−0.005	−0.004	−0.004	−0.092	−0.040	−0.199	−0.004	
		0.000	0.000	0.000	0.000	0.150	−0.002	−0.002	−0.046	−0.020	−0.100	−0.002	
		0.000	0.000	0.000	0.000	−0.003	0.148	−0.002	−0.046	−0.020	−0.100	−0.002	
		0.000	0.000	0.000	0.000	−0.003	−0.002	0.149	−0.046	−0.020	−0.100	−0.002	
		0.000	0.000	0.000	0.000	−0.003	−0.002	−0.002	0.118	−0.020	−0.100	−0.002	
		0.000	0.000	0.000	0.000	−0.003	−0.002	−0.002	−0.046	0.181	−0.100	−0.002	
		0.000	0.000	0.000	0.000	−0.003	−0.002	−0.002	−0.046	−0.020	0.104	−0.002	
		0.000	0.000	0.000	0.000	−0.003	−0.002	−0.002	−0.046	−0.020	−0.100	0.146	

(continued)

Table B.1—Continued

	Maize	Rice	Wheat	Sorghum	Barley	Other cereals	Millet	Oats	Potatoes	Sweet potatoes	Cassava
Madagascar											
Maize	0.107	–0.061	0.000						–0.010	–0.007	–0.013
Rice	–0.007	0.121	0.000						–0.010	–0.007	–0.013
Wheat	–0.007	–0.061	0.148						–0.010	–0.007	–0.013
Potatoes	–0.007	–0.061	0.000						0.143	–0.007	–0.013
Sweet potatoes	–0.007	–0.061	0.000						–0.010	0.145	–0.013
Cassava	–0.007	–0.061	0.000						–0.010	–0.007	0.140
Beans	–0.007	–0.061	0.000						–0.010	–0.007	–0.013
Peas	–0.007	–0.061	0.000						–0.010	–0.007	–0.013
Groundnuts	–0.013	–0.122	–0.001						–0.021	–0.014	–0.025
Other oil crops	–0.013	–0.122	–0.001						–0.021	–0.014	–0.025
Raw sugar	–0.007	–0.061	0.000						–0.010	–0.007	–0.013
Vegetables	–0.007	–0.061	0.000						–0.010	–0.007	–0.013
Bananas	–0.007	–0.061	0.000						–0.010	–0.007	–0.013
Fruits	–0.007	–0.061	0.000						–0.010	–0.007	–0.013
Coffee	–0.007	–0.061	0.000						–0.010	–0.007	–0.013
Spices and beverages	–0.007	–0.061	0.000						–0.010	–0.007	–0.013
Cotton lint	–0.007	–0.061	0.000						–0.010	–0.007	–0.013
Rwanda											
Maize	0.155	–0.007		–0.006			–0.001		–0.004	–0.032	–0.018
Rice	–0.001	0.144		–0.006			–0.001		–0.004	–0.032	–0.018
Sorghum	–0.001	–0.007		0.160			–0.001		–0.004	–0.032	–0.018
Millet	–0.001	–0.007		–0.006			0.149		–0.004	–0.032	–0.018
Potatoes	–0.001	–0.007		–0.006			–0.001		0.148	–0.032	–0.018
Sweet potatoes	–0.001	–0.007		–0.006			–0.001		–0.004	0.137	–0.018
Cassava	–0.001	–0.007		–0.006			–0.001		–0.004	–0.032	0.144
Other roots	–0.001	–0.007		–0.006			–0.001		–0.004	–0.032	–0.018
Beans	–0.001	–0.007		–0.006			–0.001		–0.004	–0.032	–0.018
Peas	–0.001	–0.007		–0.006			–0.001		–0.004	–0.032	–0.018
Groundnuts	–0.002	–0.014		–0.013			–0.001		–0.007	–0.065	–0.036
Other oil crops	–0.002	–0.014		–0.013			–0.001		–0.007	–0.065	–0.036
Vegetables	–0.001	–0.007		–0.006			–0.001		–0.004	–0.032	–0.018
Bananas	–0.001	–0.007		–0.006			–0.001		–0.004	–0.032	–0.018
Fruits	–0.001	–0.007		–0.006			–0.001		–0.004	–0.032	–0.018
Coffee	–0.001	–0.007		–0.006			–0.001		–0.004	–0.032	–0.018
Tea	–0.001	–0.007		–0.006			–0.001		–0.004	–0.032	–0.018
Spices and beverages	–0.001	–0.007		–0.006			–0.001		–0.004	–0.032	–0.018

Other roots	Beans	Peas	Groundnuts	Sesame seeds	Other oil crops	Raw sugar	Vegetables	Bananas	Fruits	Coffee	Tea	Spices and beverages	Cotton lint
	−0.002	0.000	−0.002		−0.001	0.000	−0.003	0.000	0.000	0.000		−0.008	−0.001
	−0.002	0.000	−0.002		−0.001	0.000	−0.003	0.000	0.000	0.000		−0.008	−0.001
	−0.002	0.000	−0.002		−0.001	0.000	−0.003	0.000	0.000	0.000		−0.008	−0.001
	−0.002	0.000	−0.002		−0.001	0.000	−0.003	0.000	0.000	0.000		−0.008	−0.001
	−0.002	0.000	−0.002		−0.001	0.000	−0.003	0.000	0.000	0.000		−0.008	−0.001
	−0.002	0.000	−0.002		−0.001	0.000	−0.003	0.000	0.000	0.000		−0.008	−0.001
	0.130	0.000	−0.002		−0.001	0.000	−0.003	0.000	0.000	0.000		−0.008	−0.001
	−0.002	0.146	−0.002		−0.001	0.000	−0.003	0.000	0.000	0.000		−0.008	−0.001
	−0.004	−0.001	0.284		−0.001	0.000	−0.006	−0.001	−0.001	0.000		−0.016	−0.001
	−0.004	−0.001	−0.004		0.298	0.000	−0.006	−0.001	−0.001	0.000		−0.016	−0.001
	−0.002	0.000	−0.002		−0.001	0.148	−0.003	0.000	0.000	0.000		−0.008	−0.001
	−0.002	0.000	−0.002		−0.001	0.000	0.142	0.000	0.000	0.000		−0.008	−0.001
	−0.002	0.000	−0.002		−0.001	0.000	−0.003	0.148	0.000	0.000		−0.008	−0.001
	−0.002	0.000	−0.002		−0.001	0.000	−0.003	0.000	0.148	0.000		−0.008	−0.001
	−0.002	0.000	−0.002		−0.001	0.000	−0.003	0.000	0.000	0.148		−0.008	−0.001
	−0.002	0.000	−0.002		−0.001	0.000	−0.003	0.000	0.000	0.000		0.164	−0.001
	−0.002	0.000	−0.002		−0.001	0.000	−0.003	0.000	0.000	0.000		−0.008	0.147
−0.003	−0.012	−0.001	−0.001		−0.001		−0.001	−0.104	−0.001	−0.001	−0.001	0.000	
−0.003	−0.012	−0.001	−0.001		−0.001		−0.001	−0.104	−0.001	−0.001	−0.001	0.000	
−0.003	−0.012	−0.001	−0.001		−0.001		−0.001	−0.104	−0.001	−0.001	−0.001	0.000	
−0.003	−0.012	−0.001	−0.001		−0.001		−0.001	−0.104	−0.001	−0.001	−0.001	0.000	
−0.003	−0.012	−0.001	−0.001		−0.001		−0.001	−0.104	−0.001	−0.001	−0.001	0.000	
−0.003	−0.012	−0.001	−0.001		−0.001		−0.001	−0.104	−0.001	−0.001	−0.001	0.000	
−0.003	−0.012	−0.001	−0.001		−0.001		−0.001	−0.104	−0.001	−0.001	−0.001	0.000	
0.150	−0.012	−0.001	−0.001		−0.001		−0.001	−0.104	−0.001	−0.001	−0.001	0.000	
−0.003	0.165	−0.001	−0.001		−0.001		−0.001	−0.104	−0.001	−0.001	−0.001	0.000	
−0.003	−0.012	0.150	−0.001		−0.001		−0.001	−0.104	−0.001	−0.001	−0.001	0.000	
−0.005	−0.024	−0.001	0.298		−0.002		−0.003	−0.208	−0.003	−0.002	−0.002	0.000	
−0.005	−0.024	−0.001	−0.002		0.304		−0.003	−0.208	−0.003	−0.002	−0.002	0.000	
−0.003	−0.012	−0.001	−0.001		−0.001		0.150	−0.104	−0.001	−0.001	−0.001	0.000	
−0.003	−0.012	−0.001	−0.001		−0.001		−0.001	0.103	−0.001	−0.001	−0.001	0.000	
−0.003	−0.012	−0.001	−0.001		−0.001		−0.001	−0.104	0.148	−0.001	−0.001	0.000	
−0.003	−0.012	−0.001	−0.001		−0.001		−0.001	−0.104	−0.001	0.149	−0.001	0.000	
−0.003	−0.012	−0.001	−0.001		−0.001		−0.001	−0.104	−0.001	−0.001	0.148	0.000	
−0.003	−0.012	−0.001	−0.001		−0.001		−0.001	−0.104	−0.001	−0.001	−0.001	0.148	

(continued)

Table B.1—Continued

	Maize	Rice	Wheat	Sorghum	Barley	Other cereals	Millet	Oats	Potatoes	Sweet potatoes	Cassava
Sudan											
Maize	0.148									0.000	0.000
Sweet potatoes	–0.001									0.149	0.000
Cassava	–0.001									0.000	0.149
Other roots	–0.001									0.000	0.000
Beans	–0.001									0.000	0.000
Peas	–0.001									0.000	0.000
Groundnuts	–0.002									–0.001	–0.001
Sesame seeds	–0.002									–0.001	–0.001
Other oil crops	–0.002									–0.001	–0.001
Vegetables	–0.001									0.000	0.000
Fruits	–0.001									0.000	0.000
Spices and beverages	–0.001									0.000	0.000
Cotton lint	–0.001									0.000	0.000
Tanzania											
Maize	0.154	–0.027	–0.031	–0.001	–0.001		–0.007		–0.011	–0.011	–0.014
Rice	–0.001	0.114	–0.031	–0.001	–0.001		–0.007		–0.011	–0.011	–0.014
Wheat	–0.001	–0.027	0.113	–0.001	–0.001		–0.007		–0.011	–0.011	–0.014
Sorghum	–0.001	–0.027	–0.031	0.151	–0.001		–0.007		–0.011	–0.011	–0.014
Barley	–0.001	–0.027	–0.031	–0.001	0.147		–0.007		–0.011	–0.011	–0.014
Millet	–0.001	–0.027	–0.031	–0.001	–0.001		0.144		–0.011	–0.011	–0.014
Potatoes	–0.001	–0.027	–0.031	–0.001	–0.001		–0.007		0.133	–0.011	–0.014
Sweet potatoes	–0.001	–0.027	–0.031	–0.001	–0.001		–0.007		–0.011	0.139	–0.014
Cassava	–0.001	–0.027	–0.031	–0.001	–0.001		–0.007		–0.011	–0.011	0.133
Other roots	–0.001	–0.027	–0.031	–0.001	–0.001		–0.007		–0.011	–0.011	–0.014
Beans	–0.001	–0.027	–0.031	–0.001	–0.001		–0.007		–0.011	–0.011	–0.014
Peas	–0.001	–0.027	–0.031	–0.001	–0.001		–0.007		–0.011	–0.011	–0.014
Groundnuts	–0.001	–0.055	–0.062	–0.002	–0.001		–0.015		–0.023	–0.021	–0.028
Sesame seeds	–0.001	–0.055	–0.062	–0.002	–0.001		–0.015		–0.023	–0.021	–0.028
Other oil crops	–0.001	–0.055	–0.062	–0.002	–0.001		–0.015		–0.023	–0.021	–0.028
Raw sugar	–0.001	–0.027	–0.031	–0.001	–0.001		–0.007		–0.011	–0.011	–0.014
Vegetables	–0.001	–0.027	–0.031	–0.001	–0.001		–0.007		–0.011	–0.011	–0.014
Bananas	–0.001	–0.027	–0.031	–0.001	–0.001		–0.007		–0.011	–0.011	–0.014
Fruits	–0.001	–0.027	–0.031	–0.001	–0.001		–0.007		–0.011	–0.011	–0.014
Coffee	–0.001	–0.027	–0.031	–0.001	–0.001		–0.007		–0.011	–0.011	–0.014
Tea	–0.001	–0.027	–0.031	–0.001	–0.001		–0.007		–0.011	–0.011	–0.014
Spices and beverages	–0.001	–0.027	–0.031	–0.001	–0.001		–0.007		–0.011	–0.011	–0.014
Cotton lint	–0.001	–0.027	–0.031	–0.001	–0.001		–0.007		–0.011	–0.011	–0.014

Other roots	Beans	Peas	Groundnuts	Sesame seeds	Other oil crops	Raw sugar	Vegetables	Bananas	Fruits	Coffee	Tea	Spices and beverages	Cotton lint
−0.001	−0.002	−0.007	−0.039	−0.005	−0.001		−0.008		−0.012			−0.001	−0.003
−0.001	−0.002	−0.007	−0.039	−0.005	−0.001		−0.008		−0.012			−0.001	−0.003
−0.001	−0.002	−0.007	−0.039	−0.005	−0.001		−0.008		−0.012			−0.001	−0.003
0.142	−0.002	−0.007	−0.039	−0.005	−0.001		−0.008		−0.012			−0.001	−0.003
−0.001	0.155	−0.007	−0.039	−0.005	−0.001		−0.008		−0.012			−0.001	−0.003
−0.001	−0.002	0.154	−0.039	−0.005	−0.001		−0.008		−0.012			−0.001	−0.003
−0.002	−0.005	−0.014	0.157	−0.009	−0.002		−0.017		−0.023			−0.002	−0.006
−0.002	−0.005	−0.014	−0.079	0.126	−0.002		−0.017		−0.023			−0.002	−0.006
−0.002	−0.005	−0.014	−0.079	−0.009	0.296		−0.017		−0.023			−0.002	−0.006
−0.001	−0.002	−0.007	−0.039	−0.005	−0.001		0.168		−0.012			−0.001	−0.003
−0.001	−0.002	−0.007	−0.039	−0.005	−0.001		−0.008		0.182			−0.001	−0.003
−0.001	−0.002	−0.007	−0.039	−0.005	−0.001		−0.008		−0.012			0.151	−0.003
−0.001	−0.002	−0.007	−0.039	−0.005	−0.001		−0.008		−0.012			−0.001	0.150
0.000	−0.037	−0.004	−0.006	−0.001	−0.007	−0.001	0.000	−0.003	−0.014	−0.007	−0.004	−0.002	0.000
0.000	−0.037	−0.004	−0.006	−0.001	−0.007	−0.001	0.000	−0.003	−0.014	−0.007	−0.004	−0.002	0.000
0.000	−0.037	−0.004	−0.006	−0.001	−0.007	−0.001	0.000	−0.003	−0.014	−0.007	−0.004	−0.002	0.000
0.000	−0.037	−0.004	−0.006	−0.001	−0.007	−0.001	0.000	−0.003	−0.014	−0.007	−0.004	−0.002	0.000
0.000	−0.037	−0.004	−0.006	−0.001	−0.007	−0.001	0.000	−0.003	−0.014	−0.007	−0.004	−0.002	0.000
0.000	−0.037	−0.004	−0.006	−0.001	−0.007	−0.001	0.000	−0.003	−0.014	−0.007	−0.004	−0.002	0.000
0.000	−0.037	−0.004	−0.006	−0.001	−0.007	−0.001	0.000	−0.003	−0.014	−0.007	−0.004	−0.002	0.000
0.000	−0.037	−0.004	−0.006	−0.001	−0.007	−0.001	0.000	−0.003	−0.014	−0.007	−0.004	−0.002	0.000
0.000	−0.037	−0.004	−0.006	−0.001	−0.007	−0.001	0.000	−0.003	−0.014	−0.007	−0.004	−0.002	0.000
0.148	−0.037	−0.004	−0.006	−0.001	−0.007	−0.001	0.000	−0.003	−0.014	−0.007	−0.004	−0.002	0.000
0.000	0.117	−0.004	−0.006	−0.001	−0.007	−0.001	0.000	−0.003	−0.014	−0.007	−0.004	−0.002	0.000
0.000	−0.037	0.149	−0.006	−0.001	−0.007	−0.001	0.000	−0.003	−0.014	−0.007	−0.004	−0.002	0.000
0.000	−0.074	−0.008	0.291	−0.003	−0.014	−0.002	0.000	−0.005	−0.027	−0.014	−0.008	−0.004	0.000
0.000	−0.074	−0.008	−0.012	0.298	−0.014	−0.002	0.000	−0.005	−0.027	−0.014	−0.008	−0.004	0.000
0.000	−0.074	−0.008	−0.012	−0.003	0.426	−0.002	0.000	−0.005	−0.027	−0.014	−0.008	−0.004	0.000
0.000	−0.037	−0.004	−0.006	−0.001	−0.007	0.147	0.000	−0.003	−0.014	−0.007	−0.004	−0.002	0.000
0.000	−0.037	−0.004	−0.006	−0.001	−0.007	−0.001	0.148	−0.003	−0.014	−0.007	−0.004	−0.002	0.000
0.000	−0.037	−0.004	−0.006	−0.001	−0.007	−0.001	0.000	0.146	−0.014	−0.007	−0.004	−0.002	0.000
0.000	−0.037	−0.004	−0.006	−0.001	−0.007	−0.001	0.000	−0.003	0.130	−0.007	−0.004	−0.002	0.000
0.000	−0.037	−0.004	−0.006	−0.001	−0.007	−0.001	0.000	−0.003	−0.014	0.140	−0.004	−0.002	0.000
0.000	−0.037	−0.004	−0.006	−0.001	−0.007	−0.001	0.000	−0.003	−0.014	−0.007	0.142	−0.002	0.000
0.000	−0.037	−0.004	−0.006	−0.001	−0.007	−0.001	0.000	−0.003	−0.014	−0.007	−0.004	0.146	0.000
0.000	−0.037	−0.004	−0.006	−0.001	−0.007	−0.001	0.000	−0.003	−0.014	−0.007	−0.004	−0.002	0.149

(continued)

Table B.1—Continued

	Maize	Rice	Wheat	Sorghum	Barley	Other cereals	Millet	Oats	Potatoes	Sweet potatoes	Cassava
Uganda											
Maize	0.157	−0.006	0.000	−0.009			−0.007		−0.002	−0.004	−0.072
Rice	−0.008	0.145	0.000	−0.009			−0.007		−0.002	−0.004	−0.072
Wheat	−0.008	−0.006	0.148	−0.009			−0.007		−0.002	−0.004	−0.072
Sorghum	−0.008	−0.006	0.000	0.152			−0.007		−0.002	−0.004	−0.072
Millet	−0.008	−0.006	0.000	−0.009			0.159		−0.002	−0.004	−0.072
Potatoes	−0.008	−0.006	0.000	−0.009			−0.007		0.147	−0.004	−0.072
Sweet potatoes	−0.008	−0.006	0.000	−0.009			−0.007		−0.002	0.148	−0.072
Cassava	−0.008	−0.006	0.000	−0.009			−0.007		−0.002	−0.004	0.097
Beans	−0.008	−0.006	0.000	−0.009			−0.007		−0.002	−0.004	−0.072
Peas	−0.008	−0.006	0.000	−0.009			−0.007		−0.002	−0.004	−0.072
Groundnuts	−0.017	−0.012	0.000	−0.017			−0.013		−0.004	−0.008	−0.145
Sesame seeds	−0.017	−0.012	0.000	−0.017			−0.013		−0.004	−0.008	−0.145
Other oil crops	−0.017	−0.012	0.000	−0.017			−0.013		−0.004	−0.008	−0.145
Raw sugar	−0.008	−0.006	0.000	−0.009			−0.007		−0.002	−0.004	−0.072
Vegetables	−0.008	−0.006	0.000	−0.009			−0.007		−0.002	−0.004	−0.072
Bananas	−0.008	−0.006	0.000	−0.009			−0.007		−0.002	−0.004	−0.072
Spices and beverages	−0.008	−0.006	0.000	−0.009			−0.007		−0.002	−0.004	−0.072
Cotton lint	−0.008	−0.006	0.000	−0.009			−0.007		−0.002	−0.004	−0.072

Demand Functions

If data are available, demand functions for each commodity can be separately defined for rural and urban areas at either the national or subnational level. At present the model includes an aggregate demand function for each commodity within a country. A demand function for a specific commodity depends on prices (for all the commodities, not only the own price) and national or subnational levels of per capita income. The demand function is constrained by the homogeneous assumption to satisfy the budget constraint (that is, total expenditures spent on all commodities equal the national or subnational income). Table B.4 shows demand elasticities. Demands for food and feed in ECA countries are shown in Tables B.5 and B.6, respectively.

Income Functions

National or subnational total income is endogenously determined in the model and is equal to the sum of production revenues from agricultural and nonagricultural activities. Because the model does not take intermediate input costs into account, prices for the agricultural goods are adjusted such that total agricultural revenue is close to the value of agricultural gross domestic product (GDP). Together with the two nonagricultural sectors, which represent industrial and service GDP, total income equals the country's total GDP.

Trade and Domestic Prices

As the name of the model suggests, a multiple market structure is specified. There is perfect substitution between any country's domestically and internationally produced commodities. However, transportation and other market costs distinguish trade in the domestic market from imports and exports. For example, although imported maize is assumed to be perfectly equivalent to domestically produced maize in any country's

Other roots	Beans	Peas	Groundnuts	Sesame seeds	Other oil crops	Raw sugar	Vegetables	Bananas	Fruits	Coffee	Tea	Spices and beverages	Cotton lint
	−0.004	−0.003	−0.014	0.000	0.000	−0.001	−0.002	−0.002				−0.040	−0.002
	−0.004	−0.003	−0.014	0.000	0.000	−0.001	−0.002	−0.002				−0.040	−0.002
	−0.004	−0.003	−0.014	0.000	0.000	−0.001	−0.002	−0.002				−0.040	−0.002
	−0.004	−0.003	−0.014	0.000	0.000	−0.001	−0.002	−0.002				−0.040	−0.002
	−0.004	−0.003	−0.014	0.000	0.000	−0.001	−0.002	−0.002				−0.040	−0.002
	−0.004	−0.003	−0.014	0.000	0.000	−0.001	−0.002	−0.002				−0.040	−0.002
	−0.004	−0.003	−0.014	0.000	0.000	−0.001	−0.002	−0.002				−0.040	−0.002
	−0.004	−0.003	−0.014	0.000	0.000	−0.001	−0.002	−0.002				−0.040	−0.002
	0.176	−0.003	−0.014	0.000	0.000	−0.001	−0.002	−0.002				−0.040	−0.002
	−0.004	0.148	−0.014	0.000	0.000	−0.001	−0.002	−0.002				−0.040	−0.002
	−0.008	−0.006	0.301	0.000	0.000	−0.002	−0.004	−0.005				−0.079	−0.005
	−0.008	−0.006	−0.028	0.298	0.000	−0.002	−0.004	−0.005				−0.079	−0.005
	−0.008	−0.006	−0.028	0.000	0.297	−0.002	−0.004	−0.005				−0.079	−0.005
	−0.004	−0.003	−0.014	0.000	0.000	0.151	−0.002	−0.002				−0.040	−0.002
	−0.004	−0.003	−0.014	0.000	0.000	−0.001	0.148	−0.002				−0.040	−0.002
	−0.004	−0.003	−0.014	0.000	0.000	−0.001	−0.002	0.152				−0.040	−0.002
	−0.004	−0.003	−0.014	0.000	0.000	−0.001	−0.002	−0.002				0.106	−0.002
	−0.004	−0.003	−0.014	0.000	0.000	−0.001	−0.002	−0.002				−0.040	0.164

consumers' demand functions, maize may still not be profitable to import for a specific country if the country's domestic price is lower than the import parity price minus any transactions costs. Maize imports can only occur in this country when its domestic demand for maize grows faster than domestic supply and its local market price rises significantly. A similar situation applies to exported commodities in a country. Even though certain horticultural products in some countries in eastern and central Africa (ECA) are exportable, if their domestic production is not competitive in international markets, either due to low productivity or high transactions costs, then their exports of horticultural products will not be profitable. Only when their domestic producer prices plus market costs are lower than the export parity price of the same product do they become profitable to export. Net imports and other uses of agricultural commodities in ECA countries are shown in Tables B.7 and B.8, respectively.

In the current model, we assume an integrated national market for each country, and bilateral trade flows across subnational regions are not captured. The differences between total supply and total demand of each commodity at the national level become net trade without identifying specific trading partners. The model considers the price gap between domestic and international markets. The import parity prices are defined as border prices (CIF prices) plus transportation and other marketing costs, whereas the export parity prices are the FOB prices minus transportation and marketing cost. Price for a specific commodity is endogenously determined in a country by its domestic supply and demand if this commodity is neither imported nor exported in this country. If a commodity is imported or exported in a country, the price of the commodity is linked to its import or export parity price.

Whether that the calibrated prices for a specific commodity within a country are

Table B.2 Growth rates by commodity sectors

	Burundi	DRC	Eritrea	Ethiopia	Kenya	Madagascar	Rwanda	Sudan	Tanzania	Uganda	ECA
Maize	0.17	−0.02	−0.01	1.78	2.87	3.55	3.11	4.11	4.56	4.61	2.96
Rice	3.30	−0.02			2.01	2.96	5.04	5.10	1.02	4.69	2.46
Wheat	1.49	−0.02	0.02	1.56	2.00	1.98	3.04	4.11	4.02	2.37	2.15
Barley			1.52	1.17	0.80				1.72		1.16
Sorghum	1.45	2.32	0.02	1.98	0.66		3.54	3.18	3.50	2.49	2.74
Millet	2.11	1.50		2.80	1.58		3.64	2.45	3.41	2.96	2.79
Oats				2.80	2.21						2.76
Other cereals			−0.07	1.65							1.63
Potatoes	2.98	0.79	0.72	2.22	3.69	2.06	4.48	2.39	0.03	3.78	3.21
Sweet potatoes	3.58	0.50		2.79	2.81	2.20	5.05	1.17	3.60	3.66	3.58
Cassava	2.00	0.75			1.57	3.15	5.08	2.70	2.74	4.48	2.06
Other roots	2.18	1.25		1.92			4.79	2.44	3.55		1.95
Beans	0.53	−0.11	3.49	−3.74		1.82	4.86	5.09	3.16	5.12	1.50
Peas	0.54	0.31	1.61	2.48	0.11	3.29	4.52	5.56	2.69	3.76	2.58
Groundnuts	1.09	0.11		3.33	4.01	2.13	3.41	4.91	2.10	2.40	3.41
Other oil crops	0.07	0.12	0.98	1.40	3.74	1.53	4.64	19.54	1.23	5.21	4.42
Sesame seeds		0.33	0.11	2.32	1.09			2.59	2.11	4.25	2.83
Vegetable oil	0.09	0.05	0.98	1.40	3.74	1.52	4.65	19.56	1.23	5.19	1.42
Raw sugar	2.09	3.19		4.54	0.10	1.98	5.13	4.12	4.04	4.87	2.98
Vegetables	2.78	0.69	1.21	1.77	0.55	2.40	3.26	2.41	3.03	3.51	1.79
Bananas	2.45	0.86		1.82	0.75	3.45	2.61	3.06	3.28	3.00	2.58
Fruits	1.16	0.68	0.66	2.49	4.20	2.39	2.51	3.00	2.46	2.58	2.56
Coffee	3.03	0.00		1.48	3.02	1.95	3.49		5.07	1.85	2.11
Tea	2.37	0.98		1.45	0.12	4.04	2.06		4.04	3.85	0.96
Spices and beverages		1.40	0.47	2.85	0.65	4.23	2.57	1.59	3.95	1.47	2.57
Cotton lint	2.11	1.94		0.76	4.02	3.03		3.21	3.19	4.93	3.14
Beef	0.18	0.10	0.17	3.36	2.53	0.27	3.98	3.91	4.00	3.38	3.01
Mutton	1.50	0.10	0.83	1.52	1.71	0.28	3.95	0.76	1.82	2.89	
Poultry eggs	0.14	0.10	3.50	4.56	2.60	4.32	0.12	2.65	1.77	4.95	
Other meat	0.27	0.10	0.89	0.27	2.46	−0.01	4.10	1.96	2.96	3.05	
Milk	−0.11	0.10	0.06	5.59	5.47	0.59	4.02	1.88	4.46	5.66	4.00
Fish	0.18	3.29	4.95	4.11	2.45	1.88	2.39	2.07	0.45	4.95	2.66

higher or lower than the import or export parity prices for the same commodity is consistent with the trade data observed in a given year. If the data show net exports for maize in Uganda, for example, the model is calibrated to a situation in which the maize price faced by Ugandan farmers equals export parity price of maize (taking into account transportation and other marketing margins). However, if the data show net imports for maize in Kenya, the model must calibrate to the situation in which maize prices faced by Kenyan consumers equals the import parity price of maize. Furthermore, if the data show that there is no trade in maize in Madagascar, the model must calibrate to a situation in which the maize price in Madagascar for farmers are higher than the export parity price (that is, the country cannot export maize without lowering its maize price), but the maize price for consumers is lower than the import parity price (that is, the country cannot import maize unless the price of maize is further increased). Under this situation, the maize price in Madagascar is endogenously determined by the country's domestic supply and demand.

For each country, there are commodities exportable or importable and hence prices are exogenously linked to world prices for these commodities. But there are more commodities that are not traded; hence their prices are endogenously determined by the country's domestic demand and supply.

Even though a commodity is calibrated as nontraded in a country in the base year, with growth in population, improvement in yield, expansion in production area of this commodity, or increase in income due to growth in other sectors, a nontraded commodity can become tradable over time. It can be exportable if increased supply of it is greater than the increased demand for it over time, and hence, the price for this commodity falls to the level comparable with its export parity price. It can also be importable if demand grows too rapidly, so that domestic price for it rises to the level comparable with its import parity price.

The model can also include regional markets for certain commodities, such as maize. To include them it is necessary to have data on trade flows and price gaps for the commodities traded regionally. The calibration process is similar to the case of foreign trade in general if data are available. For example, if the data show trade in maize from Tanzania to Kenya, then the model would calibrate to a situation such that the price for maize for Tanzanian farmers was lower than that faced by Kenyan farmers, with the difference being trade margins, including transportation costs, other marketing margins, tariffs, and other trade barriers. The model does not explicitly include government account and policy instruments, which is a common treatment in a computable general equilibrium model. The market margins between domestic and border prices, however, have taken into account tariffs and other protections, which usually cause domestic prices to be higher than the border prices.

The model is solved for the period of 2004 and 2015, taking 2003 as a starting point. All endogenous variables are solved for each year, whereas the exogenous variables are updated between two conjunct years.

Mathematical Description of the Model

Both endogenous and exogenous variables are defined in Table B.9.

Supply Functions

The yield function for crops is:

$$Y_{R,Z,i,t} = \mathrm{YA}_{\mathrm{R,Z,i,t}} P_{R,Z,i,t}^{\alpha_{R,Z,i}},$$

where $Y_{R,Z,i}$ is the yield for crop i and $P_{R,i}$ is producer price for i; $\mathrm{YA}_{\mathrm{R,Z,i}}$ is the shift parameter, which depends on fertilizer and other input use, and time trend growth rate (which varies by country or subregion).

The shift coefficient in yield function can vary with technology:

$$\mathrm{YA}_{R,Z,i,t+1} = \mathrm{YA}_{\mathrm{R,Z,i,t}}(1 + g_{Y_{R,z,i}}),$$

Table B.3 Production of agricultural commodities (000s metric tons)

	Burundi	DRC	Eritrea	Ethiopia	Kenya	Madagascar	Rwanda	Sudan	Tanzania	Uganda	ECA
Maize	125.9	1184.3	12.2	2825.1	2399.6	169.7	69.7	49.0	2618.1	1092.8	10546.3
Rice	36.7	225.6		4.1	32.6	1702.7	8.7	6.2	475.4	70.4	2562.4
Wheat	8.1	9.4	22.3	1320.4	233.0	9.8	6.0	304.2	89.9	12.0	2015.1
Sorghum	66.1	57.1	130.3	1342.7	103.0	1.0	150.5	3339.6	671.6	408.8	6270.7
Barley		0.5	33.8	971.8	49.0				5.0		1060.1
Millet	10.0	34.3	23.1	317.2	48.8		3.8	564.8	231.2	591.2	1824.4
Oats				51.7	3.6		3.8				59.1
Other cereals	0.0	108.7	13.0	1677.2	16.3	842.5	0.7	3.8	254.2	36.0	2952.4
All cereals	246.8	1619.9	234.7	8510.0	2886.0	2725.6	243.3	4267.7	4345.4	2211.2	27290.5
Potatoes	25.4	89.8	39.6	388.1	896.1	289.7	673.1	15.1	247.0	473.0	3136.9
Sweet potatoes	725.3	237.4		185.2	544.7	512.2	1019.1	8.5	786.5	2407.0	6425.8
Cassava	671.7	15976.9			599.1	2442.6	629.2	10.0	7021.4	4736.6	32087.4
Other roots	94.0	399.7	86.4	3879.2	10.0	181.0	104.1	135.2	10.2	0.0	4899.9
All root crops	1516.4	16703.8	126.0	4452.5	2049.9	3425.5	2425.5	168.8	8065.1	7616.6	46550.1
Beans	236.8	122.5	2.0	135.3	314.0	76.3	199.8	32.3	261.0	450.8	1830.7
Peas	34.9	70.4	47.6	817.5	120.4	18.8	12.9	192.7	171.8	156.0	1643.0
Groundnuts (shelled equivalent)	9.0	382.8	1.5	11.2	27.3	34.8	7.3	1005.4	74.4	142.0	1695.9
Sesame seeds		4.1	3.1	22.6	10.8			280.8	39.6	95.0	456.1
Cottonseed	1.6	18.3		28.0	13.9	16.0		110.8	97.5	41.4	327.4
Grain legumes											
Coconuts					62.5	84.4			360.0		506.9
Palms	1.2	70.2				1.8			7.5		80.7

Other oil crops	0.3	50.1	15.7	230.3	47.8	23.2	13.2	51.8	56.5	155.9	414.5
Vegetable oil	3.6	223.5	4.8	57.5	38.2	20.4	0.8	211.2	96.6	20.3	849.6
Cottonseed oil	0.2	2.7		3.4	1.5	2.4	0.0	16.8	12.3	4.5	97.9
Palm oil	1.8	168.4				3.5	0.0		6.0		183.2
Coconut oil					6.3	6.1	0.0		18.6		31.0
Raw sugar	24.7	67.3		263.1	507.4	80.6	4.3	668.5	125.8	159.6	1638.2
Vegetables	239.0	439.8	28.0	606.9	1444.9	347.3	214.9	1820.6	1164.1	538.9	6844.3
Bananas	1515.3	1502.5		81.6	1090.4	279.0	2460.8	72.0	764.3	10069.0	17834.9
Fruits	84.6	931.6	3.9	151.6	879.9	598.5	59.8	1010.9	567.9	51.6	4340.4
Coffee	23.7	43.5		225.1	64.4	61.8	17.4		48.0	197.4	681.3
Tea	7.3	2.1		0.6	272.2	0.4	15.0		24.3	29.3	351.2
Spices and beverages	543.6	1065.0	61.0	724.3	488.1	101.8	473.3	19.6	2292.8	3568.5	9337.9
Cotton lint	1.2	9.3		14.6	4.4	12.6		56.6	51.8	17.2	167.5
Beef	9.2	13.5	15.8	293.1	285.2	147.4	17.9	296.4	222.9	98.6	1400.0
Mutton	3.9	22.2	11.3	64.4	57.4	8.3	3.3	261.2	39.7	29.8	501.5
Poultry eggs	9.3	18.3	3.7	74.7	111.6	81.4	3.7	74.1	78.2	63.5	518.5
Other meat	4.3	169.9	0.7	128.0	46.3	72.6	11.2	77.8	26.2	95.9	633.0
Dairy milk	30.3	5.2	65.1	1290.2	2599.2	532.0	143.3	4879.1	839.9	545.0	10929.3
Fish	11.9	203.0	7.4	95.1	194.9	137.7	7.0	53.0	332.2	249.4	1291.6
Skins and hides	2.8	6.4	4.4	70.0	67.1	22.2	3.3	103.8	52.4	19.8	352.2
Livestock meat	17.3	205.7	27.9	485.4	388.9	228.4	32.3	635.4	288.8	224.4	2534.5
Other livestock products	24.0	227.7	15.5	239.9	373.7	241.3	14.0	230.9	462.8	332.7	2162.3
Total livestock products	71.7	438.6	108.4	2015.5	3361.8	1001.6	189.7	5745.3	1591.4	1102.0	15626.1
Fruits and vegetables	323.6	1371.4	31.9	758.5	2324.9	945.9	274.7	2831.5	1732.0	590.5	11184.7

Note: Production values are averages for 1998–2002.

Table B.4 Commodity demand price elasticities

	Maize	Rice	Wheat	Sorghum	Barley	Millet	Oats	Other cereals	Potatoes	Sweet potatoes	Cassava	Other roots	Beans	Peas	Groundnuts	Sesame seeds
Burundi																
Maize	−0.530															
Rice		−0.954														
Wheat			−0.689													
Sorghum	0.000	0.000	0.000	−0.212	0.000	0.000	0.000	0.000								
Barley																
Millet						−0.530										
Oats																
Other cereals																
Potatoes									−0.514							
Sweet potatoes										−0.514						
Cassava											−0.530					
Other roots												−0.514				
Beans													−0.909			
Peas														−0.909		
Groundnuts															−0.960	
Sesame seeds																
Other oil crops																
Vegetable oil																
Raw sugar	0.001	0.001	0.001	0.001	0.001	0.001	0.001	0.001	0.001	0.001	0.001	0.001	0.001	0.001	0.001	0.001
Vegetables																−0.700
Bananas																
Fruits																
Coffee																
Tea																
Spices and beverages																
Cotton lint																
Beef	0.001	0.001	0.001	0.001	0.001	0.001	0.001	0.001	0.001	0.001	0.001	0.001	0.001	0.001	0.001	0.001
Mutton	0.001	0.001	0.001	0.001	0.001	0.001	0.001	0.001	0.001	0.001	0.001	0.001	0.001	0.001	0.001	0.001
Poultry eggs	0.001	0.001	0.001	0.001	0.001	0.001	0.001	0.001	0.001	0.001	0.001	0.001	0.001	0.001	0.001	0.001
Other meat																
Milk	0.001	0.001	0.001	0.001	0.001	0.001	0.001	0.001	0.001	0.001	0.001	0.001	0.001	0.001	0.001	0.001
Fish	0.001	0.001	0.001	0.001	0.001	0.001	0.001	0.001	0.001	0.001	0.001	0.001	0.001	0.001	0.001	0.001
Industry	0.001	0.001	0.001	0.001	0.001	0.001	0.001	0.001	0.001	0.001	0.001	0.001	0.001	0.001	0.001	0.001
Services	0.001	0.001	0.001	0.001	0.001	0.001	0.001	0.001	0.001	0.001	0.001	0.001	0.001	0.001	0.001	0.001

where g_Y is annual growth rate in yield productivity and varies by country and crop.

The area function for crops is:

$$A_{R,Z,i,t} = AA_{R,Z,i,t} \prod_j P_{R,j,t}^{\beta_{R,Z,j}},$$

where $A_{R,Z,i}$ is the area for crop i and $P_{R,1}$, $P_{R,2}$, . . . $P_{R,j}$, is the vector of producer prices; $AA_{R,Z,i}$ is the shift parameter (the trend in area).

Trends in area function are given by:

$$AA_{R,Z,i,t+1} = AA_{R,Z,i,t}(1 + g_{A_{R,Z,i}}),$$

where g_A is annual growth rate in area expansion and varies by country and crop.

The total supply of crops is:

$$S_{R,Z,i,t} = Y_{R,Z,i,t} \cdot A_{R,Z,i,t}.$$

Other oil crops	Vegetable oil	Raw sugar	Vegetables	Bananas	Fruits	Coffee	Tea	Spices and beverages	Cotton lint	Beef	Mutton	Poultry eggs	Other meat	Milk	Fish	Industry	Services
−0.960																	
	−0.960																
0.001	0.001	−1.093	0.001	0.001	0.001	0.001	0.001	0.001	0.001	0.001	0.001	0.001	0.001	0.001	0.001	0.001	0.001
				−0.700													
					−0.700												
						−0.960											
							−0.960										
0.001	0.001	0.001	0.001	0.001	0.001	0.001	0.001	0.001	0.001	−1.533	0.001	0.001	0.001	0.001	0.001	0.001	0.001
0.001	0.001	0.001	0.001	0.001	0.001	0.001	0.001	0.001	0.001	0.001	−1.041	0.001	0.001	0.001	0.001	0.001	0.001
0.001	0.001	0.001	0.001	0.001	0.001	0.001	0.001	0.001	0.001	0.001	0.001	−1.864	0.001	0.001	0.001	0.001	0.001
										0.000	0.000	0.000	−0.807	0.000	0.000		
0.001	0.001	0.001	0.001	0.001	0.001	0.001	0.001	0.001	0.001	0.001	0.001	0.001	0.001	−2.149	0.001	0.001	0.001
0.001	0.001	0.001	0.001	0.001	0.001	0.001	0.001	0.001	0.001	0.001	0.001	0.001	0.001	0.001	−1.533	0.001	0.001
0.001	0.001	0.001	0.001	0.001	0.001	0.001	0.001	0.001	0.001	0.001	0.001	0.001	0.001	0.001	0.001	−1.233	0.001
0.001	0.001	0.001	0.001	0.001	0.001	0.001	0.001	0.001	0.001	0.001	0.001	0.001	0.001	0.001	0.001	0.001	−1.530

(continued)

The supply function for livestock and nonagriculture is given by:

$$S^{LV}_{r,Z,i,t} = \mathrm{SA}^{\mathrm{LV}}_{\mathrm{R,Z,i,t}} \prod_j P^{\beta^{LV}_{R,Z,j}}_{R,j,t}.$$

The supply function for trends in livestock and nonagricultural is:

$$\mathrm{SA}^{\mathrm{LV}}_{\mathrm{R,Z,i,t+1}} = \mathrm{SA}^{\mathrm{LV}}_{\mathrm{R,Z,i,t}}(1 + g_{S_{R,Z,i}}),$$

where g_S is annual growth rate in productivity and varies by country and commodity. The variables g_Y, g_A, and g_S are exogenous in the model and are affected by the investment shocks in the scenarios.

Demand Functions

The country-level per capita is determined by:

Table B.4—Continued

	Maize	Rice	Wheat	Sorghum	Barley	Millet	Oats	Other cereals	Potatoes	Sweet potatoes	Cassava	Other roots	Beans	Peas	Groundnuts	Sesame seeds
DRC																
Maize	−0.530															
Rice		−0.954														
Wheat			−0.689													
Sorghum	0.000	0.000	0.000	−0.212	0.000	0.000	0.000	0.000								
Barley																
Millet						−0.530										
Oats																
Other cereals																
Potatoes									−0.514							
Sweet potatoes										−0.514						
Cassava											−0.530					
Other roots												−0.514				
Beans													−0.909			
Peas														−0.909		
Groundnuts															−0.960	
Sesame seeds																−0.960
Other oil crops																
Vegetable oil																
Raw sugar	0.001	0.001	0.001	0.001	0.001	0.001	0.001	0.001	0.001	0.001	0.001	0.001	0.001	0.001	0.001	0.001
Vegetables																
Bananas																
Fruits																
Coffee																
Tea																
Spices and beverages																
Cotton lint																
Beef	0.001	0.001	0.001	0.001	0.001	0.001	0.001	0.001	0.001	0.001	0.001	0.001	0.001	0.001	0.001	0.001
Mutton	0.001	0.001	0.001	0.001	0.001	0.001	0.001	0.001	0.001	0.001	0.001	0.001	0.001	0.001	0.001	0.001
Poultry eggs	0.001	0.001	0.001	0.001	0.001	0.001	0.001	0.001	0.001	0.001	0.001	0.001	0.001	0.001	0.001	0.001
Other meat																0.001
Milk	0.001	0.001	0.001	0.001	0.001	0.001	0.001	0.001	0.001	0.001	0.001	0.001	0.001	0.001	0.001	
Fish	0.001	0.001	0.001	0.001	0.001	0.001	0.001	0.001	0.001	0.001	0.001	0.001	0.001	0.001	0.001	0.001
Industry	0.001	0.001	0.001	0.001	0.001	0.001	0.001	0.001	0.001	0.001	0.001	0.001	0.001	0.001	0.001	0.001
Services	0.001	0.001	0.001	0.001	0.001	0.001	0.001	0.001	0.001	0.001	0.001	0.001	0.001	0.001	0.001	0.001

$$Dpc_{R,Z,i,t} = \prod_j PC_{R,j,t}^{\varepsilon_{R,Z,j}} GDPpc_{R,Z,t}^{1-\Sigma_j e_{R,Z,j}},$$

where $Dpc_{R,Z,i}$ is per capita demand for commodity i in country R and subregion Z, and $PC_{R,i}$ is consumer price for commodity i in country R, $j = 1, 2, \ldots, 33$ (including two aggregate nonagricultural goods). The variable $GDPpc_{R,Z}$ is per capita agricultural and nonagricultural income for country R and subregion Z.

Foreign and Regional Trade

We assume that there are domestic market margins between import parity prices and consumer prices and between export parity prices and producer prices. Moreover, with the data available, it is possible to have different market margins across subregions within a country.

Imports and import parity prices are determined by:

Other oil crops	Vegetable oil	Raw sugar	Vegetables	Bananas	Fruits	Coffee	Tea	Spices and beverages	Cotton lint	Beef	Mutton	Poultry eggs	Other meat	Milk	Fish	Industry	Services
−0.960																	
	−0.960																
0.001	0.001	−1.093	0.001	0.001	0.001	0.001	0.001	0.001	0.001	0.001	0.001	0.001	0.001	0.001	0.001	0.001	0.001
			−0.700														
				−0.700													
					−0.700												
						−0.960											
							−0.960										
								−0.960									
0.001	0.001	0.001	0.001	0.001	0.001	0.001	0.001	0.001	0.001	−1.533	0.001	0.001	0.001	0.001	0.001	0.001	0.001
0.001	0.001	0.001	0.001	0.001	0.001	0.001	0.001	0.001	0.001	0.001	−1.041	0.001	0.001	0.001	0.001	0.001	0.001
0.001	0.001	0.001	0.001	0.001	0.001	0.001	0.001	0.001	0.001	0.001	0.001	−1.864	0.001	0.001	0.001	0.001	0.001
0.001	0.001	0.001	0.001							0.000	0.000	0.000	−0.807	0.000	0.000		
				0.001	0.001	0.001	0.001	0.001	0.001	0.001	0.001	0.001	0.001	−2.149	0.001	0.001	0.001
0.001	0.001	0.001	0.001	0.001	0.001	0.001	0.001	0.001	0.001	0.001	0.001	0.001	0.001	0.001	−1.533	0.001	0.001
0.001	0.001	0.001	0.001	0.001	0.001	0.001	0.001	0.001	0.001	0.001	0.001	0.001	0.001	0.001	0.001	−1.233	0.001
0.001	0.001	0.001	0.001	0.001	0.001	0.001	0.001	0.001	0.001	0.001	0.001	0.001	0.001	0.001	0.001	0.001	−1.945

(continued)

$$PC_{R,i,t} \leq (1 + Wmt_{R,i}) \cdot PWM_{R,i},$$

where *Wmt* is the marketing margins between country's CIF prices and consumer prices. If commodity *i* is importable, that is, $M_{R,i,t} > 0$, then this equation holds with equality.

Exports and export parity prices are given by:

$$P_{r,i,t}(1 - Wmt_{R,i}) \cdot PWE_{R,i},$$

where *P* is producer price and *PWE* is the FOB price, and the equation holds with equality when $E_{R,i,t} > 0$, that is, when commodity *i* is exportable in country *R*.

There are marketing margins between producer and consumer prices:

$$PC_{R,i,t} = (1 + DMT_{R,i}) \cdot P_{R,i,t},$$

Table B.4—Continued

	Maize	Rice	Wheat	Sorghum	Barley	Millet	Oats	Other cereals	Potatoes	Sweet potatoes	Cassava	Other roots	Beans	Peas	Groundnuts	Sesame seeds
Eritrea																
Maize	−0.530															
Rice																
Wheat			−0.689													
Sorghum	0.000	0.000	0.000	−0.212	0.000	0.000	0.000	0.000								
Barley	0.000	0.000	0.000	0.000	−0.318	0.000	0.000	0.000								
Millet																
Oats																
Other cereals								−0.954								
Potatoes									−0.514							
Sweet potatoes																
Cassava																
Other roots																
Beans													−0.909			
Peas														−0.909		
Groundnuts																
Sesame seeds																−0.960
Other oil crops																
Vegetable oil																
Raw sugar																
Vegetables																
Bananas																
Fruits																
Coffee																
Tea																
Spices and beverages																
Cotton lint																
Beef	0.001	0.001	0.001	0.001	0.001	0.001	0.001	0.001	0.001	0.001	0.001	0.001	0.001	0.001	0.001	0.001
Mutton	0.001	0.001	0.001	0.001	0.001	0.001	0.001	0.001	0.001	0.001	0.001	0.001	0.001	0.001	0.001	0.001
Poultry eggs	0.001	0.001	0.001	0.001	0.001	0.001	0.001	0.001	0.001	0.001	0.001	0.001	0.001	0.001	0.001	0.001
Other meat																
Milk	0.001	0.001	0.001	0.001	0.001	0.001	0.001	0.001	0.001	0.001	0.001	0.001	0.001	0.001	0.001	0.001
Fish	0.001	0.001	0.001	0.001	0.001	0.001	0.001	0.001	0.001	0.001	0.001	0.001	0.001	0.001	0.001	0.001
Industry	0.001	0.001	0.001	0.001	0.001	0.001	0.001	0.001	0.001	0.001	0.001	0.001	0.001	0.001	0.001	0.001
Services	0.001	0.001	0.001	0.001	0.001	0.001	0.001	0.001	0.001	0.001	0.001	0.001	0.001	0.001	0.001	0.001

where *Dmt* is the domestic marketing margin between producer and consumer prices and varies by country.

If there is regional trade for a commodity, then the following relationship holds between producer prices in the two traded countries:

$$P_{R,i,t} \geq (1 - Rmt_{R,S,i}) \cdot PS,i,t.$$

The equality holds if country R exports commodity i to country S, that is, $REGT_{R,S,i,t} > 0$.

Balance of Demand and Supply

At the national level, the balance of demand and supply is given by:

$$\sum_Z S_{R,Z,i,t} + M_{i,t} - E_{i,t} - \sum_S REGT_{R,S,i,t}$$
$$= \sum_Z Dpc_{R,Z,i,t} \cdot PoP_{R,Z}.$$

Other oil crops	Vegetable oil	Raw sugar	Vegetables	Bananas	Fruits	Coffee	Tea	Spices and beverages	Cotton lint	Beef	Mutton	Poultry eggs	Other meat	Milk	Fish	Industry	Services
−0.960																	
	−0.960																
			−0.700														
					−0.700												
								−0.960									
0.001	0.001	0.001	0.001	0.001	0.001	0.001	0.001	0.001	0.001	−1.533	0.001	0.001	0.001	0.001	0.001	0.001	0.001
0.001	0.001	0.001	0.001	0.001	0.001	0.001	0.001	0.001	0.001	0.001	−1.041	0.001	0.001	0.001	0.001	0.001	0.001
0.001	0.001	0.001	0.001	0.001	0.001	0.001	0.001	0.001	0.001	0.001	0.001	−1.864	0.001	0.001	0.001	0.001	0.001
										0.000	0.000	0.000	−0.807	0.000	0.000		
0.001	0.001	0.001	0.001	0.001	0.001	0.001	0.001	0.001	0.001	0.001	0.001	0.001	0.001	−2.149	0.001	0.001	0.001
0.001	0.001	0.001	0.001	0.001	0.001	0.001	0.001	0.001	0.001	0.001	0.001	0.001	0.001	0.001	−1.533	0.001	0.001
0.001	0.001	0.001	0.001	0.001	0.001	0.001	0.001	0.001	0.001	0.001	0.001	0.001	0.001	0.001	0.001	−1.233	0.001
0.001	0.001	0.001	0.001	0.001	0.001	0.001	0.001	0.001	0.001	0.001	0.001	0.001	0.001	0.001	0.001	0.001	−1.071

(continued)

GDP and Income Functions

The GDP function at the national level is:

$$GDP_{R,t} = \sum_j P_{R,j,t} \cdot S_{R,j,t} \; j = 1, 2, \ldots 33$$
(including nonagriculture).

The national-level per capita income function is:

$$GDPpc_{R,t} = \frac{GDP_{R,t}}{PoP_{R,t}}.$$

Table B.4—Continued

	Maize	Rice	Wheat	Sorghum	Barley	Millet	Oats	Other cereals	Potatoes	Sweet potatoes	Cassava	Other roots	Beans	Peas	Groundnuts	Sesame seeds
Ethiopia																
Maize	−0.530															
Rice																
Wheat			−0.689													
Sorghum	0.000	0.000	0.000	−0.212	0.000	0.000	0.000	0.000								
Barley	0.000	0.000	0.000	0.000	−0.318	0.000	0.000	0.000								
Millet						−0.530										
Oats							−0.530									
Other cereals								−0.954								
Potatoes									−0.514							
Sweet potatoes										−0.514						
Cassava																
Other roots												−0.514				
Beans													−0.909			
Peas														−0.909		
Groundnuts															−0.960	
Sesame seeds																−0.960
Other oil crops																
Vegetable oil																
Raw sugar	0.001	0.001	0.001	0.001	0.001	0.001	0.001	0.001	0.001	0.001	0.001	0.001	0.001	0.001	0.001	0.001
Vegetables																
Bananas																
Fruits																
Coffee																
Tea																
Spices and beverages																
Cotton lint																
Beef	0.001	0.001	0.001	0.001	0.001	0.001	0.001	0.001	0.001	0.001	0.001	0.001	0.001	0.001	0.001	0.001
Mutton	0.001	0.001	0.001	0.001	0.001	0.001	0.001	0.001	0.001	0.001	0.001	0.001	0.001	0.001	0.001	0.001
Poultry eggs	0.001	0.001	0.001	0.001	0.001	0.001	0.001	0.001	0.001	0.001	0.001	0.001	0.001	0.001	0.001	0.001
Other meat																
Milk	0.001	0.001	0.001	0.001	0.001	0.001	0.001	0.001	0.001	0.001	0.001	0.001	0.001	0.001	0.001	0.001
Fish	0.001	0.001	0.001	0.001	0.001	0.001	0.001	0.001	0.001	0.001	0.001	0.001	0.001	0.001	0.001	0.001
Industry	0.001	0.001	0.001	0.001	0.001	0.001	0.001	0.001	0.001	0.001	0.001	0.001	0.001	0.001	0.001	0.001
Services	0.001	0.001	0.001	0.001	0.001	0.001	0.001	0.001	0.001	0.001	0.001	0.001	0.001	0.001	0.001	0.001

Other oil crops	Vegetable oil	Raw sugar	Vegetables	Bananas	Fruits	Coffee	Tea	Spices and beverages	Cotton lint	Beef	Mutton	Poultry eggs	Other meat	Milk	Fish	Industry	Services
−0.960																	
	−0.960																
0.001	0.001	−1.093	0.001	0.001	0.001	0.001	0.001	0.001	0.001	0.001	0.001	0.001	0.001	0.001	0.001	0.001	0.001
			−0.700														
				−0.700													
					−0.700												
						−0.960											
							−0.960										
								−0.960									
0.001	0.001	0.001	0.001	0.001	0.001	0.001	0.001	0.001	0.001	−1.533	0.001	0.001	0.001	0.001	0.001	0.001	0.001
0.001	0.001	0.001	0.001	0.001	0.001	0.001	0.001	0.001	0.001	0.001	−1.041	0.001	0.001	0.001	0.001	0.001	0.001
0.001	0.001	0.001	0.001	0.001	0.001	0.001	0.001	0.001	0.001	0.001	0.001	−1.864	0.001	0.001	0.001	0.001	0.001
										0.000	0.000	0.000	−0.807	0.000	0.000		
0.001	0.001	0.001	0.001	0.001	0.001	0.001	0.001	0.001	0.001	0.001	0.001	0.001	0.001	−2.149	0.001	0.001	0.001
0.001	0.001	0.001	0.001	0.001	0.001	0.001	0.001	0.001	0.001	0.001	0.001	0.001	0.001	0.001	−1.533	0.001	0.001
0.001	0.001	0.001	0.001	0.001	0.001	0.001	0.001	0.001	0.001	0.001	0.001	0.001	0.001	0.001	0.001	−1.233	0.001
0.001	0.001	0.001	0.001	0.001	0.001	0.001	0.001	0.001	0.001	0.001	0.001	0.001	0.001	0.001	0.001	0.001	−1.243

(continued)

Table B.4—Continued

	Maize	Rice	Wheat	Sorghum	Barley	Millet	Oats	Other cereals	Potatoes	Sweet potatoes	Cassava	Other roots	Beans	Peas	Groundnuts	Sesame seeds
Kenya																
Maize	–0.530															
Rice		–0.954														
Wheat			–0.689													
Sorghum	0.000	0.000	0.000	–0.212	0.000	0.000	0.000	0.000								
Barley	0.000	0.000	0.000	0.000	–0.318	0.000	0.000	0.000								
Millet						–0.530										
Oats							–0.530									
Other cereals																
Potatoes									–0.514							
Sweet potatoes										–0.514						
Cassava											–0.530					
Other roots																
Beans																
Peas														–0.909		
Groundnuts															–0.960	
Sesame seeds																–0.960
Other oil crops																
Vegetable oil																
Raw sugar	0.001	0.001	0.001	0.001	0.001	0.001	0.001	0.001	0.001	0.001	0.001	0.001	0.001	0.001	0.001	0.001
Vegetables																
Bananas																
Fruits																
Coffee																
Tea																
Spices and beverages																
Cotton lint																
Beef	0.001	0.001	0.001	0.001	0.001	0.001	0.001	0.001	0.001	0.001	0.001	0.001	0.001	0.001	0.001	0.001
Mutton	0.001	0.001	0.001	0.001	0.001	0.001	0.001	0.001	0.001	0.001	0.001	0.001	0.001	0.001	0.001	0.001
Poultry eggs	0.001	0.001	0.001	0.001	0.001	0.001	0.001	0.001	0.001	0.001	0.001	0.001	0.001	0.001	0.001	0.001
Other meat																
Milk	0.001	0.001	0.001	0.001	0.001	0.001	0.001	0.001	0.001	0.001	0.001	0.001	0.001	0.001	0.001	0.001
Fish	0.001	0.001	0.001	0.001	0.001	0.001	0.001	0.001	0.001	0.001	0.001	0.001	0.001	0.001	0.001	0.001
Industry	0.001	0.001	0.001	0.001	0.001	0.001	0.001	0.001	0.001	0.001	0.001	0.001	0.001	0.001	0.001	0.001
Services																

Other oil crops	Vegetable oil	Raw sugar	Vegetables	Bananas	Fruits	Coffee	Tea	Spices and beverages	Cotton lint	Beef	Mutton	Poultry eggs	Other meat	Milk	Fish	Industry	Services
-0.960																	
	-0.960																
0.001	0.001	-1.093	0.001	0.001	0.001	0.001	0.001	0.001	0.001	0.001	0.001	0.001	0.001	0.001	0.001	0.001	0.001
			-0.700														
				-0.700													
					-0.700												
						-0.960											
							-0.960										
								-0.960									
0.001	0.001	0.001	0.001	0.001	0.001	0.001	0.001	0.001	0.001	-1.533	0.001	0.001	0.001	0.001	0.001	0.001	0.001
0.001	0.001	0.001	0.001	0.001	0.001	0.001	0.001	0.001	0.001	0.001	-1.041	0.001	0.001	0.001	0.001	0.001	0.001
0.001	0.001	0.001	0.001	0.001	0.001	0.001	0.001	0.001	0.001	0.001	0.001	-1.864	0.001	0.001	0.001	0.001	0.001
										0.000	0.000	0.000	-0.807	0.000	0.000		
0.001	0.001	0.001	0.001	0.001	0.001	0.001	0.001	0.001	0.001	0.001	0.001	0.001	0.001	-2.149	0.001	0.001	0.001
0.001	0.001	0.001	0.001	0.001	0.001	0.001	0.001	0.001	0.001	0.001	0.001	0.001	0.001	0.001	-1.533	0.001	0.001
0.001	0.001	0.001	0.001	0.001	0.001	0.001	0.001	0.001	0.001	0.001	0.001	0.001	0.001	0.001	0.001	-1.233	0.001
																	-0.921

(continued)

Table B.4—Continued

	Maize	Rice	Wheat	Sorghum	Barley	Millet	Oats	Other cereals	Potatoes	Sweet potatoes	Cassava	Other roots	Beans	Peas	Groundnuts	Sesame seeds
Madagascar																
Maize	−0.530															
Rice		−0.954														
Wheat			−0.689													
Sorghum																
Barley																
Millet																
Oats																
Other cereals																
Potatoes									−0.514							
Sweet potatoes										−0.514						
Cassava											−0.530					
Other roots																
Beans													−0.909			
Peas														−0.909		
Groundnuts															−0.960	
Sesame seeds																
Other oil crops																
Vegetable oil																
Raw sugar	0.001	0.001	0.001	0.001	0.001	0.001	0.001	0.001	0.001	0.001	0.001	0.001	0.001	0.001	0.001	0.001
Vegetables																
Bananas																
Fruits																
Coffee																
Tea																
Spices and beverages																
Cotton lint																
Beef	0.001	0.001	0.001	0.001	0.001	0.001	0.001	0.001	0.001	0.001	0.001	0.001	0.001	0.001	0.001	0.001
Mutton	0.001	0.001	0.001	0.001	0.001	0.001	0.001	0.001	0.001	0.001	0.001	0.001	0.001	0.001	0.001	0.001
Poultry eggs	0.001	0.001	0.001	0.001	0.001	0.001	0.001	0.001	0.001	0.001	0.001	0.001	0.001	0.001	0.001	0.001
Other meat																
Milk	0.001	0.001	0.001	0.001	0.001	0.001	0.001	0.001	0.001	0.001	0.001	0.001	0.001	0.001	0.001	0.001
Fish	0.001	0.001	0.001	0.001	0.001	0.001	0.001	0.001	0.001	0.001	0.001	0.001	0.001	0.001	0.001	0.001
Industry	0.001	0.001	0.001	0.001	0.001	0.001	0.001	0.001	0.001	0.001	0.001	0.001	0.001	0.001	0.001	0.001
Services																

	Other oil crops	Vegetable oil	Raw sugar	Vegetables	Bananas	Fruits	Coffee	Tea	Spices and beverages	Cotton lint	Beef	Mutton	Poultry eggs	Other meat	Milk	Fish	Industry	Services
	−0.960																	
		−0.960																
	0.001	0.001	−1.093	0.001	0.001	0.001	0.001	0.001	0.001	0.001	0.001	0.001	0.001	0.001	0.001	0.001	0.001	0.001
				−0.700														
					−0.700													
						−0.700												
							−0.960											
								−0.960										
									−0.960									
	0.001	0.001	0.001	0.001	0.001	0.001	0.001	0.001	0.001	0.001	−1.533	0.001	0.001	0.001	0.001	0.001	0.001	0.001
	0.001	0.001	0.001	0.001	0.001	0.001	0.001	0.001	0.001	0.001	0.001	−1.041	0.001	0.001	0.001	0.001	0.001	0.001
	0.001	0.001	0.001	0.001	0.001	0.001	0.001	0.001	0.001	0.001	0.001	0.001	−1.864	0.001	0.001	0.001	0.001	0.001
											0.000	0.000	0.000	−0.807	0.000	0.000		
	0.001	0.001	0.001	0.001	0.001	0.001	0.001	0.001	0.001	0.001	0.001	0.001	0.001	0.001	−2.149	0.001	0.001	0.001
	0.001	0.001	0.001	0.001	0.001	0.001	0.001	0.001	0.001	0.001	0.001	0.001	0.001	0.001	0.001	−1.533	0.001	0.001
	0.001	0.001	0.001	0.001	0.001	0.001	0.001	0.001	0.001	0.001	0.001	0.001	0.001	0.001	0.001	0.001	−1.233	0.001
																		−0.971

(continued)

Table B.4—Continued

	Maize	Rice	Wheat	Sorghum	Barley	Millet	Oats	Other cereals	Potatoes	Sweet potatoes	Cassava	Other roots	Beans	Peas	Groundnuts	Sesame seeds
Rwanda																
Maize	−0.530															
Rice		−0.954														
Wheat			−0.689													
Sorghum	0.000	0.000	0.000	−0.212	0.000	0.000	0.000	0.000								
Barley																
Millet						−0.530										
Oats																
Other cereals																
Potatoes									−0.514							
Sweet potatoes										−0.514						
Cassava											−0.530					
Other roots												−0.514				
Beans													−0.909			
Peas														−0.909		
Groundnuts															−0.960	
Sesame seeds																
Other oil crops																
Vegetable oil																
Raw sugar	0.001	0.001	0.001	0.001	0.001	0.001	0.001	0.001	0.001	0.001	0.001	0.001	0.001	0.001	0.001	0.001
Vegetables																
Bananas																
Fruits																
Coffee																
Tea																
Spices and beverages																
Cotton lint																
Beef	0.001	0.001	0.001	0.001	0.001	0.001	0.001	0.001	0.001	0.001	0.001	0.001	0.001	0.001	0.001	0.001
Mutton	0.001	0.001	0.001	0.001	0.001	0.001	0.001	0.001	0.001	0.001	0.001	0.001	0.001	0.001	0.001	0.001
Poultry eggs	0.001	0.001	0.001	0.001	0.001	0.001	0.001	0.001	0.001	0.001	0.001	0.001	0.001	0.001	0.001	0.001
Other meat																
Milk	0.001	0.001	0.001	0.001	0.001	0.001	0.001	0.001	0.001	0.001	0.001	0.001	0.001	0.001	0.001	0.001
Fish	0.001	0.001	0.001	0.001	0.001	0.001	0.001	0.001	0.001	0.001	0.001	0.001	0.001	0.001	0.001	0.001
Industry	0.001	0.001	0.001	0.001	0.001	0.001	0.001	0.001	0.001	0.001	0.001	0.001	0.001	0.001	0.001	0.001
Services	0.001	0.001	0.001	0.001	0.001	0.001	0.001	0.001	0.001	0.001	0.001	0.001	0.001	0.001	0.001	0.001

Other oil crops	Vegetable oil	Raw sugar	Vegetables	Bananas	Fruits	Coffee	Tea	Spices and beverages	Cotton lint	Beef	Mutton	Poultry eggs	Other meat	Milk	Fish	Industry	Services
−0.960																	
	−0.960																
0.001	0.001	−1.093	0.001	0.001	0.001	0.001	0.001	0.001	0.001	0.001	0.001	0.001	0.001	0.001	0.001	0.001	0.001
			−0.700														
				−0.700													
					−0.700												
						−0.960											
							−0.960										
								−0.960									
0.001	0.001	0.001	0.001	0.001	0.001	0.001	0.001	0.001	0.001	−1.533	0.001	0.001	0.001	0.001	0.001	0.001	0.001
0.001	0.001	0.001	0.001	0.001	0.001	0.001	0.001	0.001	0.001	0.001	−1.041	0.001	0.001	0.001	0.001	0.001	0.001
0.001	0.001	0.001	0.001	0.001	0.001	0.001	0.001	0.001	0.001	0.001	0.001	−1.864	0.001	0.001	0.001	0.001	0.001
										0.000	0.000	0.000	−0.807	0.000	0.000		
0.001	0.001	0.001	0.001	0.001	0.001	0.001	0.001	0.001	0.001	0.001	0.001	0.001	0.001	−2.149	0.001	0.001	0.001
0.001	0.001	0.001	0.001	0.001	0.001	0.001	0.001	0.001	0.001	0.001	0.001	0.001	0.001	0.001	−1.533	0.001	0.001
0.001	0.001	0.001	0.001	0.001	0.001	0.001	0.001	0.001	0.001	0.001	0.001	0.001	0.001	0.001	0.001	−1.233	0.001
0.001	0.001	0.001	0.001	0.001	0.001	0.001	0.001	0.001	0.001	0.001	0.001	0.001	0.001	0.001	0.001	0.001	−1.326

(continued)

Table B.4—Continued

	Maize	Rice	Wheat	Sorghum	Barley	Millet	Oats	Other cereals	Potatoes	Sweet potatoes	Cassava	Other roots	Beans	Peas	Groundnuts	Sesame seeds
Sudan																
Maize	–0.530															
Rice		–0.954														
Wheat			–0.689													
Sorghum	0.000	0.000	0.000	–0.212	0.000	0.000	0.000	0.000								
Barley																
Millet						–0.530										
Oats																
Other cereals																
Potatoes									–0.514							
Sweet potatoes										–0.514						
Cassava											–0.530					
Other roots												–0.514				
Beans													–0.909			
Peas														–0.909		
Groundnuts															–0.960	
Sesame seeds																–0.960
Other oil crops																
Vegetable oil																
Raw sugar	0.001	0.001	0.001	0.001	0.001	0.001	0.001	0.001	0.001	0.001	0.001	0.001	0.001	0.001	0.001	0.001
Vegetables																
Bananas																
Fruits																
Coffee																
Tea																
Spices and beverages																
Cotton lint																
Beef	0.001	0.001	0.001	0.001	0.001	0.001	0.001	0.001	0.001	0.001	0.001	0.001	0.001	0.001	0.001	0.001
Mutton	0.001	0.001	0.001	0.001	0.001	0.001	0.001	0.001	0.001	0.001	0.001	0.001	0.001	0.001	0.001	0.001
Poultry eggs	0.001	0.001	0.001	0.001	0.001	0.001	0.001	0.001	0.001	0.001	0.001	0.001	0.001	0.001	0.001	0.001
Other meat																
Milk	0.001	0.001	0.001	0.001	0.001	0.001	0.001	0.001	0.001	0.001	0.001	0.001	0.001	0.001	0.001	0.001
Fish	0.001	0.001	0.001	0.001	0.001	0.001	0.001	0.001	0.001	0.001	0.001	0.001	0.001	0.001	0.001	0.001
Industry	0.001	0.001	0.001	0.001	0.001	0.001	0.001	0.001	0.001	0.001	0.001	0.001	0.001	0.001	0.001	0.001
Services	0.001	0.001	0.001	0.001	0.001	0.001	0.001	0.001	0.001	0.001	0.001	0.001	0.001	0.001	0.001	0.001

	Other oil crops	Vegetable oil	Raw sugar	Vegetables	Bananas	Fruits	Coffee	Tea	Spices and beverages	Cotton lint	Beef	Mutton	Poultry eggs	Other meat	Milk	Fish	Industry	Services
	−0.960																	
		−0.960																
	0.001	0.001	−1.093	0.001	0.001	0.001	0.001	0.001	0.001	0.001	0.001	0.001	0.001	0.001	0.001	0.001	0.001	0.001
				−0.700														
					−0.700													
						−0.700												
									−0.960									
	0.001	0.001	0.001	0.001	0.001	0.001	0.001	0.001	0.001	0.001	−1.533	0.001	0.001	0.001	0.001	0.001	0.001	0.001
	0.001	0.001	0.001	0.001	0.001	0.001	0.001	0.001	0.001	0.001	0.001	−1.041	0.001	0.001	0.001	0.001	0.001	0.001
	0.001	0.001	0.001	0.001	0.001	0.001	0.001	0.001	0.001	0.001	0.001	0.001	−1.864	0.001	0.001	0.001	0.001	0.001
											0.000	0.000	0.000	−0.807	0.000	0.000		
	0.001	0.001	0.001	0.001	0.001	0.001	0.001	0.001	0.001	0.001	0.001	0.001	0.001	0.001	−1.083	0.001	0.001	0.001
	0.001	0.001	0.001	0.001	0.001	0.001	0.001	0.001	0.001	0.001	0.001	0.001	0.001	0.001	0.001	−1.533	0.001	0.001
	0.001	0.001	0.001	0.001	0.001	0.001	0.001	0.001	0.001	0.001	0.001	0.001	0.001	0.001	0.001	0.001	−1.233	0.001
	0.001	0.001	0.001	0.001	0.001	0.001	0.001	0.001	0.001	0.001	0.001	0.001	0.001	0.001	0.001	0.001	0.001	−2.917

(continued)

Table B.4—Continued

	Maize	Rice	Wheat	Sorghum	Barley	Millet	Oats	Other cereals	Potatoes	Sweet potatoes	Cassava	Other roots	Beans	Peas	Groundnuts	Sesame seeds
Tanzania																
Maize	–0.530															
Rice		–0.954														
Wheat			–0.689													
Sorghum	0.000	0.000	0.000	–0.212	0.000	0.000	0.000	0.000								
Barley	0.000	0.000	0.000	0.000	–0.318	0.000	0.000	0.000								
Millet						–0.530										
Oats																
Other cereals																
Potatoes									–0.514							
Sweet potatoes										–0.514						
Cassava											–0.530					
Other roots												–0.514				
Beans													–0.909			
Peas														–0.909		
Groundnuts															–0.960	
Sesame seeds																–0.960
Other oil crops																
Vegetable oil																
Raw sugar	0.001	0.001	0.001	0.001	0.001	0.001	0.001	0.001	0.001	0.001	0.001	0.001	0.001	0.001	0.001	0.001
Vegetables																
Bananas																
Fruits																
Coffee																
Tea																
Spices and beverages																
Cotton lint																
Beef	0.001	0.001	0.001	0.001	0.001	0.001	0.001	0.001	0.001	0.001	0.001	0.001	0.001	0.001	0.001	0.001
Mutton	0.001	0.001	0.001	0.001	0.001	0.001	0.001	0.001	0.001	0.001	0.001	0.001	0.001	0.001	0.001	0.001
Poultry eggs	0.001	0.001	0.001	0.001	0.001	0.001	0.001	0.001	0.001	0.001	0.001	0.001	0.001	0.001	0.001	0.001
Other meat																
Milk	0.001	0.001	0.001	0.001	0.001	0.001	0.001	0.001	0.001	0.001	0.001	0.001	0.001	0.001	0.001	0.001
Fish	0.001	0.001	0.001	0.001	0.001	0.001	0.001	0.001	0.001	0.001	0.001	0.001	0.001	0.001	0.001	0.001
Industry	0.001	0.001	0.001	0.001	0.001	0.001	0.001	0.001	0.001	0.001	0.001	0.001	0.001	0.001	0.001	0.001
Services	0.001	0.001	0.001	0.001	0.001	0.001	0.001	0.001	0.001	0.001	0.001	0.001	0.001	0.001	0.001	0.001

Other oil crops	Vegetable oil	Raw sugar	Vegetables	Bananas	Fruits	Coffee	Tea	Spices and beverages	Cotton lint	Beef	Mutton	Poultry eggs	Other meat	Milk	Fish	Industry	Services
−0.960																	
	−0.960																
0.001	0.001	−1.093	0.001	0.001	0.001	0.001	0.001	0.001	0.001	0.001	0.001	0.001	0.001	0.001	0.001	0.001	0.001
			−0.700														
				−0.700													
					−0.700												
						−0.960											
							−0.960										
								−0.960									
0.001	0.001	0.001	0.001	0.001	0.001	0.001	0.001	0.001	0.001	−1.533	0.001	0.001	0.001	0.001	0.001	0.001	0.001
0.001	0.001	0.001	0.001	0.001	0.001	0.001	0.001	0.001	0.001	0.001	−1.041	0.001	0.001	0.001	0.001	0.001	0.001
0.001	0.001	0.001	0.001	0.001	0.001	0.001	0.001	0.001	0.001	0.001	0.001	−1.864	0.001	0.001	0.001	0.001	0.001
										0.000	0.000	0.000	−0.807	0.000	0.000		
0.001	0.001	0.001	0.001	0.001	0.001	0.001	0.001	0.001	0.001	0.001	0.001	0.001	0.001	−2.149	0.001	0.001	0.001
0.001	0.001	0.001	0.001	0.001	0.001	0.001	0.001	0.001	0.001	0.001	0.001	0.001	0.001	0.001	−1.533	0.001	0.001
0.001	0.001	0.001	0.001	0.001	0.001	0.001	0.001	0.001	0.001	0.001	0.001	0.001	0.001	0.001	0.001	−1.233	0.001
0.001	0.001	0.001	0.001	0.001	0.001	0.001	0.001	0.001	0.001	0.001	0.001	0.001	0.001	0.001	0.001	0.001	−1.077

(continued)

Table B.4—Continued

	Maize	Rice	Wheat	Sorghum	Barley	Millet	Oats	Other cereals	Potatoes	Sweet potatoes	Cassava	Other roots	Beans	Peas	Groundnuts	Sesame seeds
Uganda																
Maize	−0.530															
Rice	−0.954															
Wheat	−0.689															
Sorghum	0.000	0.000	0.000	−0.212	0.000	0.000	0.000	0.000								
Barley																
Millet						−0.530										
Oats																
Other cereals																
Potatoes						−0.514										
Sweet potatoes						−0.514										
Cassava											−0.530					
Other roots																
Beans											−0.909					
Peas											−0.909					
Groundnuts											−0.960					
Sesame seeds																−0.960
Other oil crops																−0.960
Vegetable oil																−0.960
Raw sugar	0.001	0.001	0.001	0.001	0.001	0.001	0.001	0.001	0.001	0.001	0.001	0.001	0.001	0.001	0.001	0.001
Vegetables																−0.700
Bananas																
Fruits																
Coffee																
Tea																
Spices and beverages																
Cotton lint																
Beef	0.001	0.001	0.001	0.001	0.001	0.001	0.001	0.001	0.001	0.001	0.001	0.001	0.001	0.001	0.001	0.001
Mutton	0.001	0.001	0.001	0.001	0.001	0.001	0.001	0.001	0.001	0.001	0.001	0.001	0.001	0.001	0.001	0.001
Poultry eggs	0.001	0.001	0.001	0.001	0.001	0.001	0.001	0.001	0.001	0.001	0.001	0.001	0.001	0.001	0.001	0.001
Other meat																
Milk	0.001	0.001	0.001	0.001	0.001	0.001	0.001	0.001	0.001	0.001	0.001	0.001	0.001	0.001	0.001	0.001
Fish	0.001	0.001	0.001	0.001	0.001	0.001	0.001	0.001	0.001	0.001	0.001	0.001	0.001	0.001	0.001	0.001
Industry	0.001	0.001	0.001	0.001	0.001	0.001	0.001	0.001	0.001	0.001	0.001	0.001	0.001	0.001	0.001	0.001
Services	0.001	0.001	0.001	0.001	0.001	0.001	0.001	0.001	0.001	0.001	0.001	0.001	0.001	0.001	0.001	0.001

Notes: The elasticities employed in the study are obtained from various sources, including the research literature, expert suggestions, and educated guesses. These elasticities are then further adjusted to fit into the data (for example, the share of each crop production value in total agricultural revenue by country is used to adjust cross-price elasticity in the supply functions, and consumer budget share by country is used to derive cross-price elasticity in the demand functions). Because of this adjustment, values of many cross-price elasticities carry more than 10 decimal places. For reporting convenience, the approximated numbers with only the first three decimal places are shown in the table (blank entries thus indicate values less than 0.001).

Other oil crops	Vegetable oil	Raw sugar	Vegetables	Bananas	Fruits	Coffee	Tea	Spices and beverages	Cotton lint	Beef	Mutton	Poultry eggs	Other meat	Milk	Fish	Industry	Services
0.001	0.001	−1.093	0.001	0.001	0.001	0.001	0.001	0.001	0.001	0.001	0.001	0.001	0.001	0.001	0.001	0.001	0.001
				−0.700													
				−0.700													
				−0.960													
				−0.960													
				−0.960													
0.001	0.001	0.001	0.001	0.001	0.001	0.001	0.001	0.001	0.001	−1.533	0.001	0.001	0.001	0.001	0.001	0.001	0.001
0.001	0.001	0.001	0.001	0.001	0.001	0.001	0.001	0.001	0.001	0.001	−1.041	0.001	0.001	0.001	0.001	0.001	0.001
0.001	0.001	0.001	0.001	0.001	0.001	0.001	0.001	0.001	0.001	0.001	0.001	−1.864	0.001	0.001	0.001	0.001	0.001
									0.000	0.000	0.000	−0.807		0.000	0.000		
0.001	0.001	0.001	0.001	0.001	0.001	0.001	0.001	0.001	0.001	0.001	0.001	0.001	0.001	−2.149	0.001	0.001	0.001
0.001	0.001	0.001	0.001	0.001	0.001	0.001	0.001	0.001	0.001	0.001	0.001	0.001	0.001	0.001	−1.533	0.001	0.001
0.001	0.001	0.001	0.001	0.001	0.001	0.001	0.001	0.001	0.001	0.001	0.001	0.001	0.001	0.001	0.001	−1.233	0.001
0.001	0.001	0.001	0.001	0.001	0.001	0.001	0.001	0.001	0.001	0.001	0.001	0.001	0.001	0.001	0.001	0.001	−1.134

Table B.5 Food demand (000s metric tons)

	Burundi	DRC	Eritrea	Ethiopia	Kenya	Madagascar	Rwanda	Sudan	Tanzania	Uganda	ECA
Maize	129.9	1098.6	18.3	2727.1	2628.9	147.7	89.5	75.7	2462.4	690.1	10068.3
Rice	33.2	258.6	2.3	5.3	122.5	1494.0	21.9	35.0	596.9	100.7	2670.4
Wheat	18.1	348.5	229.3	1931.1	790.8	102.5	17.9	847.6	310.8	89.1	4685.6
Sorghum	7.9	51.4	171.7	1504.0	58.6	0.9	108.3	2977.5	286.9	125.4	5292.5
Barley	0.0	0.0	27.6	791.8	1.7		0.0	0.0	0.1	0.4	821.6
Millet	2.2	23.3	26.8	251.5	28.9		3.7	517.8	142.3	409.0	1405.6
Oats		0.2	0.0	49.9	2.8	0.0	0.9		0.0	0.0	53.7
Other cereals	34.0	6.6	14.1	1604.5	3.2	0.8	1.3	15.9	1.7	0.7	1682.7
All cereals	225.2	1787.2	490.1	8865.1	3637.5	1745.9	243.4	4469.5	3801.0	1415.3	26680.3
Potatoes	13.6	103.5	34.9	315.6	711.6	169.9	332.5	11.7	529.3	308.3	2530.8
Sweet potatoes	662.7	224.9	0.0	175.0	495.7	315.2	952.5	7.7	708.3	1859.4	5401.3
Cassava	559.3	15014.4	0.0		580.7	2091.9	512.9	9.0	6185.1	2240.0	27193.1
Other roots	93.6	350.0	82.4	3339.6	9.9	92.8	88.6	121.3	9.5	0.0	4187.7
All root crops	1329.2	15692.6	117.4	3830.2	1797.9	2669.8	1886.5	149.7	7432.2	4407.6	39313.0
Beans	226.5	113.8	2.6	105.6	8.1	67.7	181.7	20.1	215.8	369.7	1311.7
Peas	31.1	80.7	46.0	742.1	193.8	10.9	27.8	216.0	145.7	136.2	1630.3
Groundnuts (shelled equivalent)	4.0	264.4	1.6	3.6	24.4	7.2	2.2	142.1	37.1	97.4	584.1
Sesame seeds		4.0	2.0	1.0	10.4			31.5		84.5	133.3
Cottonseed		0.0	0.0	3.5						0.0	3.5
Grain legumes											
Coconuts			0.0		14.8	29.0		0.3	128.9	0.0	173.0
Palms		0.0	0.0							0.0	0.0

Other oil crops	0.6	45.8	0.0	46.7	3.0	0.2	12.1	7.3	5.4	104.9	179.2
Vegetable oil	4.4	226.1	14.7	88.6	227.2	51.8	16.0	232.7	148.9	40.8	1009.3
Cottonseed oil		2.6	0.0	3.5	1.6	6.4	0.0	23.1	13.0	3.7	139.0
Palm oil	2.2	154.5	0.0	15.9	181.9	4.3	10.6	51.0	38.9	17.7	464.6
Coconut oil	0.0	0.0	0.0	0.1	7.1	6.6	0.0	0.2	18.8	0.1	48.8
Raw sugar	25.4	132.3	11.9	249.5	603.0	125.9	19.8	496.8	248.1	183.5	1846.7
Vegetables	215.1	408.1	33.4	536.2	1347.5	309.8	184.0	968.2	997.5	480.8	5480.5
Bananas	543.3	775.9	0.0	72.8	972.9	248.6	1302.4	60.9	555.5	5040.2	9572.5
Fruits	76.1	839.2	4.0	135.2	667.2	488.1	56.5	860.2	514.8	47.0	3688.3
Coffee	0.1	6.1	0.0	106.5	7.0	28.4	0.2	9.4	2.9	12.5	173.2
Tea	0.1	1.4	0.3	1.0	18.7	0.1	0.2	20.2	4.0	5.8	51.8
Spices and beverages	496.3	1052.9	63.1	710.9	489.1	92.4	443.4	20.3	2307.0	3318.2	8993.6
Cotton lint			0.0							0.0	0.0
Beef	9.2	20.9	16.3	290.4	282.7	147.6	17.5	286.2	217.1	96.8	1384.6
Mutton	3.9	22.6	11.4	62.4	57.7	8.2	3.2	254.6	39.4	29.5	492.9
Poultry eggs	8.7	30.7	3.4	67.4	97.2	75.7	3.5	66.7	73.4	56.1	482.8
Other meat	4.3	171.7	0.7	127.9	45.4	72.3	11.1	77.3	26.2	94.5	631.4
Dairy milk	35.1	36.9	78.6	1190.4	2415.2	521.3	144.8	4716.5	820.1	483.5	10442.5
Fish	12.3	201.4	7.1	15.0	47.3	93.4	6.2	36.5	130.8	189.4	739.5
Skin and hides		0.0	0.0							0.0	0.0
Livestock meat	17.4	215.1	28.4	480.7	385.8	228.1	31.8	618.1	282.7	220.8	2508.9
Other livestock products	21.0	232.1	10.4	82.4	144.6	169.1	9.7	103.3	204.2	245.6	1222.3
Total livestock products	73.4	484.2	117.5	1753.5	2945.6	918.4	186.3	5437.9	1307.0	949.9	14173.8
Fruits and vegetables	291.2	1247.3	37.3	671.4	2014.7	797.8	240.5	1828.5	1512.3	527.8	9168.8

Note: Net import values are averages for 1998–2001.

Table B.6 Feed demand (000s metric tons)

	Burundi	DRC	Eritrea	Ethiopia	Kenya	Madagascar	Rwanda	Sudan	Tanzania	Uganda	ECA
Maize	1.3	18.8		117.5	107.5	7.9		1.6	112.5	109.1	476.2
Rice				0.0		60.0				0.0	60.0
Wheat										0.0	0.0
Sorghum					11.1			122.9	13.2	41.0	188.3
Barley								0.0		0.0	0.0
Millet					4.0			18.7	4.3	59.2	86.2
Oats					0.6				0.0	0.0	0.7
Other cereals					5.5			0.0		0.0	5.5
All cereals	1.3	18.8	0.0	117.5	128.8	67.9	0.0	143.3	130.0	209.3	816.9
Potatoes						15.8				0.0	15.8
Sweet potatoes		4.8				142.3				0.0	147.1
Cassava		918.9				221.1			35.3	1838.0	3013.2
Other roots		4.8				52.9				0.0	57.7
All root crops	0.0	928.5	0.0		0.0	432.0	0.0	0.0	35.3	1838.0	3233.8
Beans										0.0	0.0
Peas										0.0	0.0
Groundnuts (shelled equivalent)										0.0	0.0
Sesame seeds										0.0	0.0
Cottonseed										0.0	0.0
Grain legumes											
Coconuts										0.0	0.0
Palms										0.0	0.0

Other oil crops		0.1		0.1	2.1	0.0	0.0	0.1		0.0	2.3
Vegetable oil							0.0			0.0	0.1
Cottonseed oil							0.0			0.0	0.0
Palm oil							0.0			0.0	0.0
Coconut oil							0.0			0.0	0.0
Raw sugar										0.0	0.0
Vegetables										0.0	0.0
Bananas		7.0							12.3	2337.5	2356.8
Fruits										0.0	0.0
Coffee										0.0	0.0
Tea										0.0	0.0
Spices and beverages								0.1		0.0	0.1
Cotton lint				0.0						0.0	0.0
Beef										0.0	0.0
Mutton										0.0	0.0
Poultry eggs										0.0	0.0
Other meat										0.0	0.0
Dairy milk				14.3	15.8			97.8		0.0	127.9
Fish						1.7				0.0	1.7
Skins and hides										0.0	0.0
Livestock meat	0.0	0.0	0.0	0.0	0.0	0.0	0.0	0.0	0.0	0.0	0.0
Other livestock products	0.0	0.0	0.0	0.0	0.0	1.7	0.0	0.0	0.0	0.0	1.7
Total livestock products	0.0	0.0	0.0	14.3	15.8	1.7	0.0	97.8	0.0	0.0	129.5
Fruits and vegetables										0.0	0.0

Note: Feed demands are averages for 1998–2001.

Table B.7 Net imports of agricultural commodities (000s metric tons)

	Burundi	DRC	Eritrea	Ethiopia	Kenya	Madagascar	Rwanda	Sudan	Tanzania	Uganda	ECA
Maize	6.7	63.5	2.8	24.8	281.0	2.4	19.8	39.5	115.9	–1.9	554.6
Rice	3.0	57.8	2.3	5.4	92.9	130.5	16.0	31.1	138.6	41.1	518.6
Wheat	11.4	297.6	212.5	831.1	578.0	94.2	13.9	676.4	216.0	80.1	3011.2
Sorghum	0.0	0.0	22.6	28.3	–0.4		0.4	–103.4	–0.3	–0.5	–53.2
Barley	12.4	16.4	5.4	11.7	1.1	5.9	7.4	0.0	15.6	23.6	99.6
Millet		0.0	0.0	–0.2	0.6		0.1	–8.9	–0.4	–0.2	–8.9
Oats		0.2	0.0	0.1	0.3	0.0	0.9	0.0	0.0	0.0	1.5
Other cereals	3.8	6.6	2.1	–11.2	5.5	0.8	1.3	5.5	–7.1	0.0	7.4
All cereals	37.3	442.1	247.8	890.0	959.1	233.7	59.9	640.3	478.4	142.3	4130.9
Potatoes	0.1	39.2	0.0	–2.8	–0.2	0.2	0.3	0.3	7.5	0.1	44.6
Sweet potatoes		0.0	0.0	0.0					0.0	0.0	0.0
Cassava	0.0	0.0	0.0	0.0	1.1	0.1	0.1	0.0	–2.5	0.4	–0.8
Other roots	0.0	0.0	0.0	0.0	0.9	0.0	0.0	0.1	–1.4	0.0	–0.6
All root crops	0.1	39.2	0.0	–2.8	1.8	0.3	0.4	0.4	3.5	0.4	43.2
Beans	0.1	1.5	0.0	–24.2	8.1	–1.2	3.3	1.6	–8.3	–2.5	–21.5
Peas	0.0	21.0	4.1	12.9	3.1	–2.6	17.4	33.2	–10.0	–1.2	77.8
Groundnuts (shelled equivalent)	0.0	0.0	0.0	0.0	0.2	–0.4	0.0	–15.8	–0.4	0.1	–16.3
Sesame seeds		0.0	–0.9	–30.8	–0.1			–167.9	–14.3	–1.1	–215.1
Cottonseed	0.1	0.0	0.0	–0.2	0.2	0.0		–4.0	–1.3	0.7	–4.4
Grain legumes											
Coconuts			0.0		–0.3	–0.1		0.3	0.0	–0.8	–0.9
Palms	0.0	0.0	0.0		0.0	0.0			–0.3	0.0	–0.3

Other oil crops	0.3	0.1	0.0	−8.4	3.9	−0.1	0.0	−66.2	2.3	−0.4	−60.1
Vegetable oil	0.8	21.7	9.9	61.6	243.5	34.1	12.6	49.9	123.5	73.3	561.0
Cottonseed oil		0.0	0.0	0.1	0.0	3.5	0.0		1.6	−0.6	66.1
Palm oil	0.4	4.0	0.0	21.6	235.1	4.6	8.9	64.2	97.5	64.5	479.2
Coconut oil	0.0		0.0	0.1	0.7	0.5	0.0	0.2	0.3	0.1	23.5
Raw sugar	−0.5	52.0	9.2	−5.9	145.8	32.1	15.4	−105.5	119.7	52.3	320.4
Vegetables	2.5	12.4	7.0	−8.6	−50.8	−4.2	2.1	6.6	−4.3	−0.5	−37.8
Bananas	0.0	−0.1	0.0	−0.5	0.0	0.0	0.0	0.0	−0.1	−1.1	−1.7
Fruits	0.0	2.3	0.0	−0.2	−131.4	−15.9	0.2	−1.4	−2.9	0.6	−148.7
Coffee	−26.7	−24.9	0.0	−105.7	−64.3	−12.3	−12.4	9.4	−46.8	−186.2	−470.0
Tea	−6.4	−0.9	0.3	0.3	−233.7	−0.3	−14.1	20.2	−22.2	−22.5	−279.4
Spices and beverages	−1.4	2.0	0.1	−0.4	−1.8	−13.3	0.9	1.2	11.1	2.4	0.9
Cotton lint	0.1	4.9	0.0	−2.1	1.2	−0.2	0.3	−42.6	−27.9	−8.8	−75.1
Beef	0.0	7.1	0.1	0.0	−0.1	0.0	0.1	−3.1	0.0	0.0	4.2
Mutton	0.0	0.1	0.0	−1.4	0.2	0.0	0.0	−6.3	0.0	0.0	−7.4
Poultry eggs	0.0	14.1	0.0	0.0	0.0	0.1	0.0	0.2	0.7	0.1	15.3
Other meat	0.0	1.1	0.0	0.0	−0.8	0.1	0.0	0.0	0.5	0.2	1.1
Dairy milk	5.8	31.9	14.8	7.6	23.4	17.3	12.1	62.2	26.1	2.6	203.8
Fish	0.0	0.0	0.0	0.0	−0.6	−0.6	0.0	−0.1	−0.8	0.0	−2.1
Skins and hides	−0.4	0.0	−0.2	−5.7	−18.1	−2.1	−0.7	−3.8	−9.4	−18.1	−58.6
Livestock meat	0.0	8.4	0.1	−1.4	−0.7	0.1	0.1	−9.4	0.5	0.2	−2.1
Other livestock products	−0.4	14.1	−0.2	−5.7	−18.7	−2.6	−0.6	−3.7	−9.4	−18.0	−45.4
Total livestock products	5.4	54.4	14.7	0.4	4.0	14.7	11.6	49.1	17.2	−15.3	156.2
Fruits and vegetables	2.5	14.8	7.0	−8.7	−182.2	−20.1	2.2	5.2	−7.3	0.1	−186.5

Note: Net import values are averages for 1998–2001.

Table B.8 Other uses of agricultural commodities

	Burundi	DRC	Eritrea	Ethiopia	Kenya	Madagascar	Rwanda	Sudan	Tanzania	Uganda	ECA
Maize					1.7	0.0	0.0		4.8	0.3	7.3
Rice					0.4		0.0		0.0	0.1	0.5
Wheat					2.7	0.0	0.0		0.0		1.8
Sorghum											0.0
Barley											0.0
Millet											0.0
Oats											0.0
Other cereals											0.0
All cereals	0.0	0.0	0.0	0.0	4.9	0.1	0.0	0.0	4.9	0.4	9.6
Potatoes					0.9	0.2				0.0	1.1
Sweet potatoes											0.0
Cassava					1.3	108.8			55.4		167.4
Other roots											0.0
All root crops	0.00	0.00			2.3	109.0	0.0		55.4	0.0	7.3
Beans											0.0
Peas											0.0
Groundnuts (shelled equivalent)								276.3			295.0
Sesame seeds											0.0
Cottonseed											0.0
Grain legumes											
Coconuts								0.0		0.0	0.0
Palms											0.0

Other oil crops						2.3		0.4	0.0		2.3
Vegetable oil	0.4	17.6	0.1	31.4	61.6	8.3	0.8	22.2	50.9	59.3	184.6
Cottonseed oil											29.8
Palm oil	0.4	17.4		5.7	59.9	6.1	0.0	20.4	44.6	52.7	164.5
Coconut oil											6.2
Raw sugar									0.1		0.1
Vegetables	0.4	0.1									0.4
Bananas									0.0	0.0	0.0
Fruits					0.0	13.7	0.0		0.0		13.7
Coffee		15.3					4.3				19.4
Tea							0.7				0.8
Spices and beverages	0.0				10.7	1.1	0.0	0.1		0.2	11.7
Cotton lint	1.3	14.1		12.6	5.4	13.3	0.3	12.3	28.8	6.2	70.8
Beef								0.0			0.0
Mutton								0.0			0.0
Poultry eggs								0.0			0.0
Other meat								0.0			0.0
Dairy milk				0.1	1.6			0.0			0.9
Fish											0.0
Skins and hides	2.5	6.3	3.9	58.3	62.9	13.9	3.0	91.0	42.3	7.1	194.7
Livestock meat											0.0
Other livestock products											194.7
Total livestock products											195.7
Fruits and vegetables											14.2

Note: Use values are averages for 1998–2001.

Table B.9. Variables used in the economywide multimarket model

Variable	Definition
Endogenous variables solved for year t	
$Y_{R,Z,i,t}$	Crop yield
$A_{R,Z,i,t}$	Crop area
$S_{R,Z,i,t}$	Output
$Dpc_{R,i,t}$	Demand per capita
$E_{R,i,t}$	Exports
$M_{R,i,t}$	Imports
$REGT_{R,S,i,t}$	Regional net exports
$P_{R,i,t}$	Producer price
$PC_{R,i,t}$	Consumer price
$GDP_{R,t}$	Gross domestic product
$GDPpc_{R,t}$	Per capita income
Exogenous variables updated from year t to t+1 based on growth rate	
$\mathrm{YA}_{R,Z,i,t}$	Level of productivity in yield function
$\mathrm{AA}_{R,Z,i,t}$	Area level in area function
$\mathrm{SA}^{LV}_{R,Z,i,t}$	Level of productivity in supply function for noncrops
$PoP_{R,t}$	Population
Exogenous variables fixed for year t	
$PWE_{R,i}$	Export border price
$PWM_{R,i}$	Import border price
$Wmt_{R,i}$	Market margins for imports or exports
$Rmt_{R,i}$	Market margins for regional trade
$Dmt_{R,i}$	Domestic market margins
$g_{Y,R,i}$	Yield growth rate
$g_{A,R,i}$	Area growth rate
$g^{LV}_{S,R,i}$	Productivity growth rate in noncrop production
$g_{PoP,r}$	Population growth rate

APPENDIX C

The Dynamic Research Evaluation for Management Model

The Dynamic Research Evaluation for Management Model (DREAM; Alston, Norton, and Pardey 1998) was designed to measure returns to commodity-oriented research in an open-economy setting, allowing for price and technology spillover effects between a country in which the research originates and the rest of the world. DREAM is a single-commodity model, so there is no explicit representation of cross-commodity substitution effects in production and consumption, but these aspects are represented implicitly by the elasticities of supply and demand.

The primary parameterization of the supply and demand equations relies on a set of prices and annual quantities in a defined base period and a set of corresponding price elasticities. The idea is that the linear approximation implied by these elasticities will be good for small equilibrium displacements, such as those implied by single-digit percentage shifts of supply or demand, regardless of the true (nonlinear) functional forms of supply and demand. Small shifts have the added virtue that the cross-commodity and general equilibrium effects are likely to be small, and the total research benefits will not depend significantly on the particular elasticity values used (although the distribution of those total benefits among producers and consumers is sensitive to the elasticity values used).

DREAM parameterization defines the supply and demand curves in the base year to replicate observed market prices and quantities. DREAM also allows for underlying growth of supply and demand to project a stream of shifting supply and demand curves into the future that generates a stream of changing equilibrium prices and quantities, in the "without-research" scenario. These without-research outcomes can be compared to "with-research" outcomes, in which a stream of supply curve displacements also incorporates research-induced supply shifts. The research-induced supply shifts are defined by combining an estimate of a maximum percentage research-induced supply shift that could be achieved if the innovation is adopted with an assumed profile of the likely levels of adoption over time.

Finally, measures of producer and consumer surplus are computed and compared between the with-research and without-research scenarios, and these are discounted back to the base year to compute present values of benefits. In a situation where we have estimates of the costs of the research that is responsible for the supply shift being modeled, we can compute a net present value or internal rate of return, but that is not done in this study; the work here is limited to computing the present value of benefits from 1 percent supply shifts of various sorts.

The DREAM software that implements the model (Wood, You, and Baitx 2001a,b) is designed to facilitate the adjustment of parameters defining the size and time path of the research-induced supply shifts, the underlying growth rates, and the elasticities that define the slopes of the curves. Thus it is straightforward to test a range of scenarios and conduct sensitivity analyses. In this report, we have used an extension of the DREAM framework related to the spatial dimension, because we are modeling a large geographic region, within which space matters for trade because of transportation costs and other trade barriers. Specifically, we incorporate a spatial element by introducing price transmission elasticities with values less than one between countries, which thus dampen the transmission of price signals. It is a crude treatment, as we have only one transmission elasticity between each country and all others in the region, but it does have the effect of suppressing the cross-country and cross-regional price and quantity responses to changes arising in a particular country.[31]

DREAM assesses the present value of research benefits in cases with

- multiple regions,
- a homogeneous product,
- linear supply and demand in each region,
- exponential (parallel) exogenous growth of linear supply and demand,
- a parallel research-induced supply shift in one region (or multiple regions),
- a consequent parallel research-induced supply shift in other regions,
- a range of market-distorting policies,
- zero transport costs (at least initially),
- a research lag followed by a linear adoption curve up to a maximum, or
- an eventual linear decline.

The analytical model described in detail below is embedded in the DREAM computer program (Wood, You, and Baitx 2001a,b), developed for research priority setting and evaluation.

General Form of Supply and Demand

For region i in year t, linear supply-and-demand equations for a particular commodity (subscript suppressed) are specified as:

$$\text{Supply: } Q_{i,t} = a_{it} + \beta_i PP_{i,t} \qquad (1a)$$

$$\text{Demand: } C_{i,t} = \gamma_{it} + \delta_i PC_{i,t}, \qquad (1b)$$

where the first subscript, i, refers to a region, and the second subscript, t, refers to years from the initial starting point of the evaluation. The slopes are assumed to be constant for each region for all time periods. The intercepts may grow over time to reflect underlying growth in supply or demand due to factors other than research (that is, growth in productivity or income).

Initial Parameterization

Supply and demand are defined by initial conditions ($t = 0$) for the variables

- quantity consumed in each region $C_{i,0}$,
- quantity produced in each region $Q_{i,0}$,
- producer price in each region $PP_{i,0}$,
- consumer price in each region $PC_{i,0}$,
- elasticity of supply in each region $\varepsilon_{i,0}$, and
- elasticity of demand in each region $\eta_{i,0}$ (<0).

In many cases, the initial values of elasticities are assumed to be equal among regions (a convenient, but not necessary, assumption). These initial values are sufficient to allow us to compute the slope and intercept of supply and demand in each region for the initial year:

$$\beta_{i0} = \varepsilon_{i0} Q_{i,0} / PP_{i,0} \qquad (2a)$$

$$\alpha_{i0} = (1 - \varepsilon_{i0}) Q_{i,0} \qquad (2b)$$

$$\delta_{i0} = \eta_{i0} C_{i,0} / PC_{i,0} \qquad (2c)$$

$$\gamma_{i0} = (1 - \eta_{i0}) C_{i,0}. \qquad (2d)$$

Exogenous Growth in Supply and Demand

We incorporate average exponential growth rates for demand (stemming from growth in

[31]DREAM allows for the explicit inclusion of price wedges between each region and a nominal base region, to reflect structural price differences (reflecting different transport and transaction costs, price policies, and the like).

population and income) and supply (from growth in productivity or an increase in area cropped) expected to occur regardless of whether the research program is undertaken:

$$\alpha_{it} = \alpha_{it-1} + \pi_i^Q Q_{i,t} \quad \text{for } t > 0 \tag{3a}$$

$$\gamma_{it} = \gamma_{it-1} + \pi_i^C C_{i,t} \quad \text{for } t > 0, \tag{3b}$$

where π_i^C = the growth rate of demand (population growth rate + income elasticity × income growth rate) and π_i^Q = the growth rate of supply (area growth rate + yield growth rate not attributable to research).

Now we have sufficient information to parameterize the supply-and-demand equations for each region in each year under the without-research scenario.

Research-Induced Supply Shifts

Local Effect of Research. Let region i undertake a program of research with probability of success p_i, which, if the research is successful and the results are fully adopted, will yield a cost saving per unit of output equal to c_i percent of the initial price, $PP_{i,0}$ in region i, while a ceiling adoption rate of A_i^{MAX} percent holds in region i. Then it is anticipated that the supply function in region i will eventually shift down (in the price direction) by an amount per unit equal to:

$$k_i^{MAX} = p_i c_i A_i^{MAX} PP_{i,0} \geq 0. \tag{4}$$

The actual supply shift in any particular year is some fraction of the eventual maximum supply shift, k_i^{MAX}, defined in equation 4. To define the actual supply shift, we can combine the maximum supply shift with other information about the shape of the time path of $k_{i,t}$ based on data about adoption and depreciation-cum-obsolescence factors. Assuming a trapezoidal shape for the adoption curve, to define the entire profile of supply shifts over time, we need to define the parameters

- research lag in years λ_R,
- adoption lag (years from initial adoption to maximum adoption) λ_A,
- maximum lag (years from maximum adoption to eventual decline) λ_M, and
- decline lag (years from the beginning to the end of the decline) λ_D.

We can now define the supply shifts (in the price direction) for region i in each year t:

$$k_{i,t} = 0 \quad (\text{for } 0 \leq t \leq \lambda_R)$$

$$k_{i,t} = k_i^{MAX}(t - \lambda_R)/\lambda_A \quad (\text{for } \lambda_R < t \leq \lambda_R + \lambda_A)$$

$$k_{i,t} = k_i^{MAX} \quad (\text{for } \lambda_R + \lambda_A < t \leq \lambda_R + \lambda_A + \lambda_M)$$

$$k_{i,t} = k_i^{MAX} \frac{\lambda_R + \lambda_A + \lambda_M + \lambda_D - t}{\lambda_D}$$
$$(\text{for } \lambda_R + \lambda_A + \lambda_M < t \leq \lambda_R + \lambda_A + \lambda_M + \lambda_D)$$

$$k_{i,t} = 0 \quad (\text{for } t > \lambda_R + \lambda_A + \lambda_M + \lambda_D).$$

Figure C.1b shows the trapezoidal adoption curve and shows how the parameters above (λ_R, λ_A, λ_M, and λ_D) may be used to define the entire curve.

Spillover Effects of Research. The spillover effects from region i to region j are parameterized in relation to the supply shifts in region i, implicitly assuming the same adoption curve applies in every region:

$$k_{j,t} = \theta_{ji} k_{i,t} \quad \text{for all } i \text{ and } j, \tag{5}$$

where θ_{ji} = the supply shift in j stemming from research-induced supply shift in i (θ_{ii} = 1).

With-Research Supply and Demand

To model the with-research case (denoted by superscript R on all relevant variables and parameters), we take the intercepts from the without-research case (but include the effects of exogenous supply growth), add the

Figure C.1 Key analytical components of DREAM

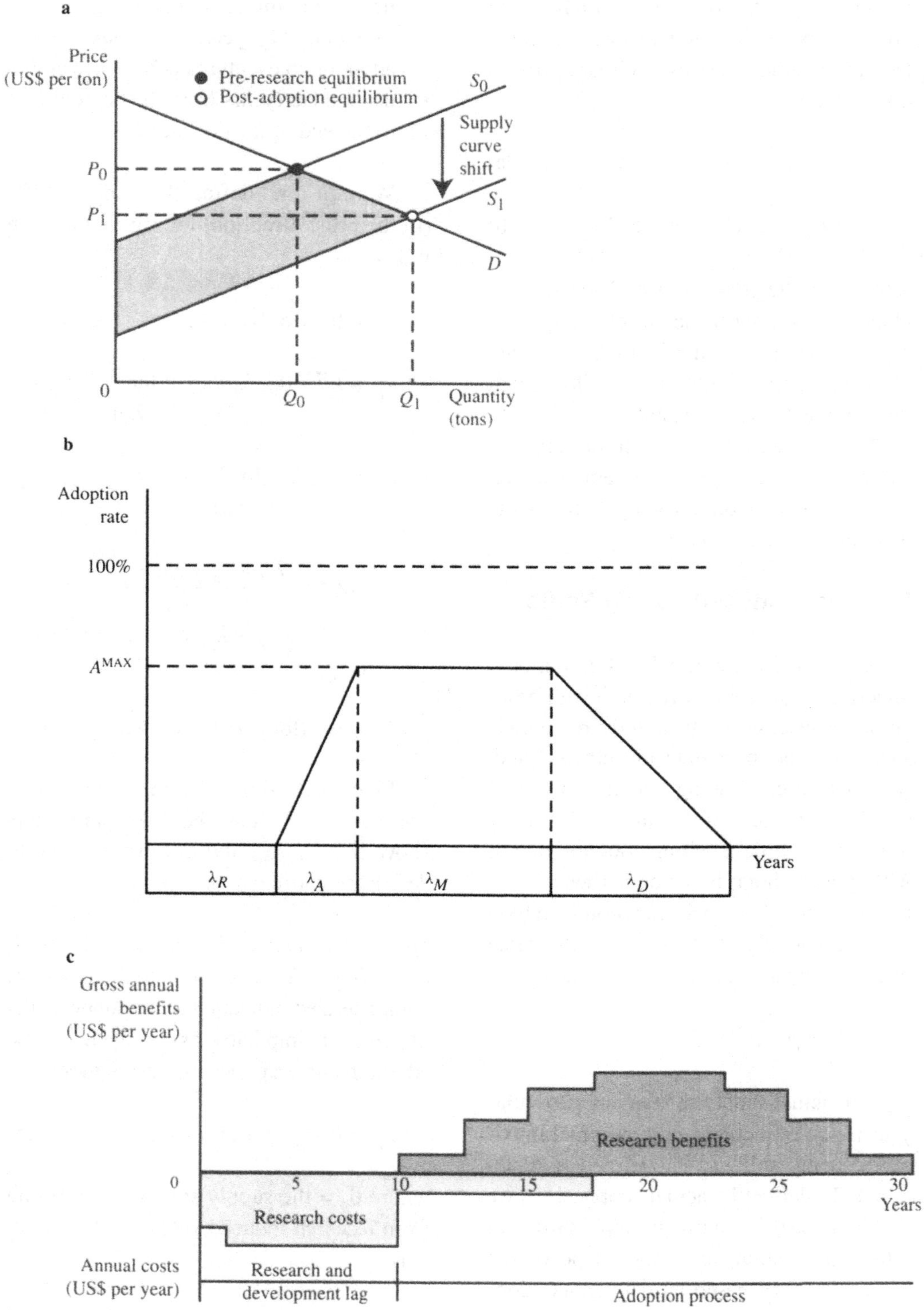

effect of the supply shift to them, and include the result in the supply equation:

$$\alpha^R_{j,t} = \alpha_{jt} + k_{jt}\beta_j. \tag{6}$$

Supply-and-demand models reflecting the local and spillover effects of research are:

$$Q^R_{i,t} = \alpha^R_{it} + \beta_i PP^R_{i,t} \tag{7a}$$

$$C^R_{i,t} = \gamma_{it} + \delta_i PC^R_{i,t} \tag{7b}$$

The only substantive difference from the corresponding without-research equations (1a and 1b) is in the supply intercept, but as noted above, the prices and quantities are labeled differently (the R superscript) to distinguish them from the without-research values:

- quantity consumed in each region $C^R_{i,t}$,
- quantity produced in each region $Q^R_{i,t}$,
- producer price in each region $PP^R_{i,t}$, and
- consumer price in each region $PC^R_{i,t}$.

Market-Clearing Rules

For all scenarios considered, there is an overall quantity-clearing rule: the sum of quantities supplied equals the sum of quantities demanded in each year. Considering n regions,

$$\begin{aligned} Q_t &= (Q_{1,t} + Q_{2,t} + \ldots + Q_{n,t}) \\ &= C_t = (C_{1,t} + C_{2,t} + \ldots + C_{n,t}). \end{aligned} \tag{8}$$

All of the market-clearing rules express policies in terms of price wedges that permit differences between consumer and producer prices within and among regions consistent with clearing quantities produced and consumed.[32]

Free Trade. The easiest case is that of free trade, where with-research prices,

$$PP^R_{i,t} = PC^R_{i,t} = PC^R_{j,t} = PP^R_{j,t} = P^R_t,$$

and without-research prices,

$$PP_{i,t} = PC_{i,t} = PC_{j,t} = PP_{j,t} = P_t,$$

are defined for all regions i and j and for any year t.

Making this substitution into each of the n regional supply-and-demand equations and then substituting them into equation 8 yields a solution for the equilibrium price for each year. To simplify, let us define the following aggregated parameters for each year t:

$$\gamma_t = \gamma_{1t} + \gamma_{2t} + \ldots + \gamma_{nt}$$

$$\alpha_t = \alpha_{1t} + \alpha_{2t} + \ldots + \alpha_{nt}$$

$$\alpha^R_t = \alpha^R_{1t} + \alpha^R_{2t} + \ldots + \alpha^R_{nt}$$

$$\delta_t = \delta = \delta_{10} + \delta_{20} + \ldots + \delta_{n0} < 0$$

$$\beta_t = \beta = \beta_{10} + \beta_{20} + \ldots + \beta_{n0} > 0.$$

Then the without-research and the with-research market-clearing prices under free trade are given by:

$$P_t = (\gamma_t + \alpha_t)/(\beta - \delta) \tag{9a}$$

$$P^R_t = (\gamma_t - \alpha^R_t)/(\beta - \delta). \tag{9b}$$

These are always positive numbers, with $P_t > P^R_t$, because the intercepts on the quantity axis satisfy $\gamma_t > \alpha^R_t > \alpha_t$, unless we make a mistake, such as letting supply grow too fast relative to demand.

We can substitute the results for prices from equations 9a and 9b into the regional supply-and-demand equations to compute regional quantities produced and consumed with and without research and then calculate

[32]Transportation costs influence trade among countries and should theoretically be incorporated into the analysis. However, accurate calculation of these costs is often difficult, because it requires knowing the transportation differentials for each commodity between the home country being studied and each of its major trading partners, as well as the pattern of commodity flows.

the regional consumer and producer welfare effects.

Generalized Taxes and Subsidies. We can define a general solution for a large variety of tax or subsidy regimes by setting out a general model in which a per unit tax is collected from consumers in every region and from producers in every region, where T_i^C = the per unit consumer tax in region i and T_i^Q = the per unit producer tax in region i.

Different policies can be represented as different combinations of taxes and subsidies:

- consumption tax in region i at T_i per unit: $T_i^C = T_i$; $T_i^Q = 0$,
- production tax in region i at T_i per unit: $T_i^C = 0$; $T_i^Q = T_i$,
- export tax in region i at T_i per unit: $T_i^C = -T_i$; $T_i^Q = T_i$,
- import tariff in region i at T_i per unit: $T_i^C = T_i$; $T_i^Q = -T_i$.

A subsidy is a negative tax, so it is also possible to use these to represent subsidies on output, consumption, imports, or exports. Suppose there is a region with no taxes or subsidies in which the prices to producers and consumers are $P_t = PC_t = PP_t$ and $P_t^R = PC_t^R = PP_i^R$. Then P_t (expressed in common currency units, either local currency or U.S. dollars) is the border price for an exporter or an importer whose internal consumer or producer prices will be equal to that price in the absence of any domestic distortions. The arbitrage rules are that the prices in all regions are:

$$PP_{i,t} = P_t - T_i^Q$$

$$PC_{i,t} = P_t + T_i^C$$

$$PP_{i,t}^R = P_t^R - T_i^Q$$

$$PC_{i,t}^R = P_t^R + T_i^C,$$

for all regions i and j and for any year t.

Making these substitutions into each of the n regional supply-and-demand equations and substituting them into equation 9 yields a solution for the equilibrium price for each year. As for the case of free trade, define the following aggregated parameters for each year:

$$\gamma_t = \gamma_{1t} + \gamma_{2t} + \ldots + \gamma_{nt}$$

$$\alpha_t = \alpha_{1t} + \alpha_{2t} + \ldots + \alpha_{nt}$$

$$\alpha_t^R = \alpha_{1t}^R + \alpha_{2t}^R + \ldots + \alpha_{nt}^R$$

$$\delta_t = \delta = \delta_{10} + \delta_{20} + \ldots + \delta_{n0} < 0$$

$$\beta_t = \beta = \beta_{10} + \beta_{20} + \ldots + \beta_{n0} > 0.$$

In addition, we define the following aggregated demand-and-supply shifts in the quantity direction because of consumer and producer taxes:

$$T_t^C = T_{1t}^C\delta_{10} + T_{2t}^C\delta_{20} + \ldots + T_{nt}^C\delta_{n0}$$

$$T_t^Q = T_{1t}^Q\beta_{10} + T_{2t}^Q\beta_{20} + \ldots + T_{nt}^Q\beta_{n0}$$

$$P_t = (\gamma_t + T_t^Q + T_t^C - \alpha_t)/(\beta - \delta) \quad (10a)$$

$$P_t^R = (\gamma_t + T_t^Q + T_t^C - \alpha_t^R)/(\beta - \delta) \quad (10b)$$

To compute the actual consumer and producer prices in any region, the results of equations 10a and 10b are substituted into the arbitrage (market-clearing) rules given above. Individual prices can then be used in the individual supply-and-demand equations (equations 1 and 7) to compute quantities with and without research, and then to compute surplus effects. Notice that this set of results includes the free-trade model as a special case (that is, when all of the taxes and subsidies are zero).

Other Policies. Quantitative restrictions on production or trade can be treated approximately as tax or subsidy equivalents with a little care to distribute tax revenue as quota rents. The approximation is somewhat unreliable in a dynamic model, but it might suffice for our purposes. A target price, deficiency-payment scheme might involve more work. Conceptually, the approach is to define target price and allow it to determine output in regions where it applies. Then, with that supply as exogenous, supply equations in the other regions and demand equations in all regions would interact to determine price.

Welfare Effects

The following equations apply for assessing welfare effects:

$$\Delta PS_{j,t} = (k_{j,t} + PP^R_{j,t} - PP_{j,t}) [Q_{j,t} + 0.5(Q^R_{j,t} - Q_{j,t})] \quad (11a)$$

$$\Delta CD_{j,t} = (PC_{j,t} - PC^R_{j,t}) [C_{j,t} + 0.5(C^R_{j,t} - C_{j,t})] \quad (11b)$$

$$\Delta GS_{j,t} = T^C_{jt}(C^R_{j,t} - C_{j,t}) + T^Q_{jt}(Q^R_{j,t} - Q_{j,t}), \quad (11c)$$

where $\Delta PS_{j,t}$ is the producer research benefit in region j in year t, $\Delta CS_{j,t}$ is the consumer research benefit in region j in year t, and $\Delta GS_{j,t}$ is the government research benefit in region j in year t.

Aggregation over Time and Interest Groups

The model generates a series of prices, quantities, and economic surplus measures for the regions of interest for a range of tax or subsidy policies. The remaining problem is to aggregate those measures into summary measures of research benefits. For a given policy scenario, we have the measure of benefits ($\Delta PS_{i,t}$, $\Delta CS_{i,t}$, $\Delta GS_{i,t}$) for each region in each time period.

The real discount rate must be defined for the computation of the present value of the stream of benefits. A reasonable approach is to fix a single value for all regions, interest groups, and years so that $r_{i,t} = r_{j,s} = r$.

We need to define a relevant planning horizon. Thirty years should be adequate for most purposes if we are using discount rates of 5 percent per year or greater. The present values of benefits to interest groups are then defined as:

$$VPS_i = \sum_{t=0}^{30} \Delta PS_{i,t}/(1 + r)^t$$

$$\Delta PS_{i,0} + \Delta PS_{i,1}/(1 + r) + \Delta PS_{i,2}/(1 + r)^2 + \ldots + \Delta PS_{i,30}/(1 + r)^{30} \quad (12a)$$

$$VCS_i = \sum_{t=0}^{30} \Delta CS_{i,t}/(1 + r)^t$$

$$= \Delta CS_{i,0} + \Delta CS_{i,1}/(1 + r) + \Delta CS_{i,2}/(1 + r)^2 + \ldots + \Delta CS_{i,30}/(1 + r)^{30} \quad (12b)$$

$$VGS_i = \sum_{t=0}^{30} \Delta GS_{i,t}/(1 + r)^t$$

$$= \Delta GS_{i,0} + \Delta GS_{i,1}/(1 + r) + \Delta GS_{i,2}/(1 + r)^2 + \ldots + \Delta GS_{i,30}/(1 + R)^{30}. \quad (12c)$$

DREAM Simulations: Potential Benefits from Regional Coordination

For the DREAM simulations reported in Chapter 5, we define 13 distinct geopolitical regions, 10 of which are defined as the countries of the Association for Strengthening Agricultural Research in Eastern and Central Africa (ASARECA). The rest of Sub-Saharan Africa and the rest of the world are two additional independent regions, with the former including all the countries of western and southern Africa, whereas the latter includes all countries and regions outside Sub-Saharan Africa. The inclusion of the last two regions into DREAM is to allow for different trade conditions for different commodities (see Table C.1 for maize). The extent to which a commodity is traded in regional, international, or only domestic markets influences the size of price effects from research-induced supply shifts. There are numerous agricultural commodities (both crop and livestock) produced in ASARECA member countries. However, only a small

Table C.1 DREAM baseline data: Maize

Region	Country	Supply (tons)	Demand (tons)	Price (US$ per ton)	Elasticity: Supply	Elasticity: Demand	Demand growth variables: Income elasticity	Demand growth variables: GDP per capita growth (% per year)
ECA	Burundi	126,125	120,221	144.45	1.0	0.5	0.51	1.03
	DRC	1,199,446	1,244,514	170.00	1.0	0.5	0.51	-3.59
	Eritrea	19,629	19,629	144.45	1.0	0.5	0.50	0.59
	Ethiopia	2,778,502	2,504,152	175.00	1.0	0.5	0.52	–0.20
	Kenya	2,300,000	2,242,270	190.72	1.0	0.5	0.52	2.35
	Madagascar	161,000	162,090	110.98	1.0	0.5	0.53	–2.38
	Rwanda	58,677	167,109	144.45	1.0	0.5	0.50	–0.16
	Sudan	44,000	76,784	82.03	1.0	0.5	0.51	0.96
	Tanzania	2,562,487	2,678,674	241.00	1.0	0.5	0.52	0.67
	Uganda	1,024,333	1,020,046	199.00	1.0	0.5	0.56	2.55
	ECA total	10,274,200	10,235,487					
Rest of Sub-Saharan Africa		24,588,213	23,558,875	126.10	1.0	0.5	0.63	0.58
Rest of world		569,563,524	570,611,905	112.19	1.0	0.5	0.87	1.36
World total		604,425,937	604,406,267					

Notes: All data are 2001–03 averages except GDP per capita growth rate, which is the 1990–2003 average annual growth rate. Prices (unit values) are calculated as export value divided by export quantity. If the price cannot be reliably calculated, the price for that country is set to be equal to the regional average for Sub-Saharan Africa (US$144.45 per ton for maize).

number of commodities dominate production and consumption in the region, especially those produced by the majority smallholder farm population. Based on the shares of both production and consumption, potential productivity growth opportunities, and market and trade opportunities, we selected the following 15 commodities for the analysis: plantain, maize, cassava, sorghum, potatoes and sweet potatoes, rice, cow milk, dry beans, groundnuts, vegetables, beef, coffee, mutton, cotton, cashew nuts. Among the commodities analyzed, cashew nuts, coffee, cotton, dry beans, maize, rice, vegetables, and beef are all considered as internationally traded, whereas cassava, groundnuts, and potatoes are assumed to be traded within the region, and plantains, potatoes, sorghum, millet, cow milk, and mutton within domestic markets only. For commodities traded in domestic markets only, countries are treated as having a closed economy.

Baseline data requirements, similar to those illustrated for maize in Table C.1, are needed for each commodity.[33] For such crops as maize that are used for both food and feed, food/feed ratios are used to compute the income elasticities. This calculation is done by weighting demand elasticity in terms of the fraction of the consumption consumed as food (rather than feed), f_i. For example, assuming a higher income elasticity of demand for feed (1.0) than for food (0.5), the income elasticity for commodity i is simply derived as $1 - 0.5f_i$. For other crops, income elasticity estimates from secondary sources in the literature are used. Demand growth over the simulation period was estimated on the basis of projected national population and national income growth rates (UNPD 1999). In the simulation, exogenous production growth is assumed to be equal to demand growth to maintain constant real prices in the absence of technical change.

[33]The full set of parameters is available from IFPRI upon request.

References

Abdulai, A., X. Diao, and M. Johnson. 2005. *Achieving regional growth dynamics in African agriculture.* DSGD Discussion Paper 17. Washington, D.C.: International Food Policy Research Institute.

Alston, J. M. 2002. Spillovers. *Australian Journal of Agricultural and Resource Economics* 46 (3): 315–346.

Alston, J. M., G. W. Norton, and P. G. Pardey. 1998. *Science under scarcity: Principles and practice for agricultural research evaluation and priority setting.* Wallingford: CAB International.

APDB (African Policy Dialogues on Biotechnology). 2004. African Policy Dialogues on Biotechnology —Building consensus and strategies for common action. <http://ifpri.org/africadialogue/index.htm>.

ASARECA (Association for Strengthening Agricultural Research in Eastern and Central Africa). 1997. *Regional collaboration in agricultural research: Development of a long-term strategic plan for regional agricultural research in the eastern and central Africa sub-region.* Entebbe: ASARECA.

Boserup, E. 1965. *The conditions of agricultural growth: The economics of agrarian change under population pressure.* London: Allen and Unwin.

———. 1981. *Population and technological change.* Chicago: University of Chicago Press.

Chema, S., J. Roseboom, and E. Gilbert. 2004. *A review of key issues and recent experiences with reforming agricultural research in Africa.* ISNAR Research Report 24. The Hague: International Service for National Agricultural Research.

CIESIN, IFPRI, World Bank, and CIAT (Center for International Earth Science Information Network, International Food Policy Research Institute, World Bank, and Centro Internacional de Agricultura Tropical). 2004. *Global Rural-Urban Mapping Project (GRUMP): Gridded population of the world,* version 3, with urban reallocation (GPW-UR). Palisades, N.Y., U.S.A.: CIESIN, Columbia University.

Delgado, C. L., J. Hopkins, V. A. Kelly, with P. Hazell, A. A. McKenna, P. Gruhn, B. Hojjati, J. Sil, and C. Courbois. 1998. *Agricultural growth linkages in Sub-Saharan Africa.* Research Report 107. Washington, D.C.: International Food Policy Research Institute.

Diao, X., M. Johnson, S. Gavian, and P. Hazell. 2005. *Africa without borders: Building blocks for regional growth.* IFPRI Issue Brief. Washington, D.C.: International Food Policy Research Institute.

EAC (East African Community). 2004. East African Community. <http://www.eac.int>.

EARRNET (Eastern Africa Rootcrops Research Network). 2003. *Setting priorities in ASARECA's networks, programs and projects—EARRNET.* Kampala: EARRNET.

ECAPAPA (Eastern and Central Africa Program for Agricultural Policy Analysis). 2002. *Harmonization of seed policies and regulations in eastern Africa: Results and agreements.* ECAPAPA Monograph Series 4. Entebbe: ECAPAPA.

———. 2003. *Setting priorities in ASARECA's networks, programs and projects—ECAPAPA.* Entebbe: ECAPAPA.

Ehui, S., and J. Pender. 2005. Resource degradation, low agricultural productivity, and poverty in Sub-Saharan Africa: Pathways out of the spiral. In *Reshaping agriculture's contributions to society: Proceedings of the twenty-fifth International Conference of Agricultural Economists,* ed. D. Colman and N. Vink. Ames, Iowa: Blackwell.

FAO (Food and Agriculture Organization of the United Nations). No date. FAOSTAT database, various years. <http://faostat.fao.org>.

———. 2002. *The state of food insecurity in the world 2002.* Rome: FAO.

Farrington, J. 1998. *Organisational roles in farmer participatory research and extension: Lessons from the last decade.* ODI Natural Resource Perspectives 27. London: Overseas Development Institute.

Fischer, G., M. Shah, H. van Velthuizen, and F. O. Nachtergaele. 2001. *Global agro-ecological assessment for agriculture in the 21st century.* Laxenburg, Austria: International Institute for Applied Systems Analysis. <http://www.iiasa.ac.at/Research/LUC/Papers/gaea.pdf>.

Gabre-Madhin, E. 2001. *Market institutions, transactions costs, and social capital in the Ethiopian grain market.* Research Report 124. Washington, D.C.: International Food Policy Research Institute.

Global Vegetation Monitoring. 2004. Unit Global Land Cover 2000 Project and datasets (GLC 2000). <http://www.gvm.sai.jrc.it/glc2000/defaultGLC2000.htm>.

Haggblade, S., P. Hazell, and T. Reardon. 2002. Strategies for stimulating poverty-alleviating growth in the rural nonfarm economy in developing countries. EPTD Discussion Paper 92. Washington, D.C.: International Food Policy Research Institute.

Holloway, G. J., and S. Ehui. 2001. Demand, supply and willingness to pay for extension services in an emerging-market setting. *American Journal of Agricultural Economics* 83 (3): 764–768.

IGAD (Inter-Governmental Authority for Development). 2004. Political and Human Affairs Division. <http://www.igad.org>.

Jayne, T. S., J. Govereh, A. Mwanaumo, J. K. Nyoro, and A. Chapoto. 2002. False promise or false premise? The experience of food and input market reform in eastern and southern Africa. *World Development* 30 (11): 1967–1985.

Jones, W. O. 1972. *Marketing staple food crops in tropical Africa.* Ithaca, N.Y., U.S.A.: Cornell University Press.

KACE (Kenya Agricultural Commodity Exchange). 2005. Kenya Agricultural Commodity Exchange. <http://www.kacekenya.com/home/index.asp>.

Kherallah, M., C. Delgado, E. Gabre-Madhin, N. Minot, and M. Johnson. 2002. Reforming agricultural markets in Africa. Baltimore: Johns Hopkins University Press.

Kilambya, D., S. Nandwa, and S. W. Omamo. 1998. Priority setting in a factor-based research program. In *Agricultural research priority setting: Information investments for improved use of research resources,* ed. B. F. Mills. The Hague: International Service for National Agricultural Research.

Lanjouw, P., and G. Feder. 2001. Rural non-farm activities and rural development: From experience towards strategy. Rural Strategy Background Paper 4. Washington, D.C.: World Bank, Rural Development Department.

Mills, B. F., ed. 1998. *Agricultural research priority setting: information investments for improved use of research resources.* The Hague: International Service for National Agricultural Research.

Mills, B. F., and D. Karanja. 1997. Processes and methods for research programme priority setting: The experience of the Kenya Agricultural Research Institute Wheat Programme. *Food Policy* 22 (1): 63–79.

NAADS (Uganda National Agricultural Advisory Services). 2004. Uganda National Agricultural Advisory Services. <www.naads.or.ug>.

NARO (National Agricultural Research Organization). 2000. *NARO medium term plan, 2001–2005.* Entebbe: NARO.

National Imagery and Mapping Agency. 1997. Vector map level 0 (digital chart of the world), 3rd ed. Stock Number VMAP0DATA. Fairfax, Va., U.S.A.: National Imagery and Mapping Agency.

NEPAD (New Partnership for Africa's Development). 2004. *Comprehensive Africa Agriculture Development Programme.* Pretoria: NEPAD.

New, M., M. Hulme, and P. Jones. 2000. Representing twentieth-century space–time climate variability. Part II: Development of 1901–96 monthly grids of terrestrial surface climate. *Journal of Climate* 13 (7): 2217–2238.

Ngigi, M. 2004. Agricultural policy challenges and objectives in eastern and central Africa. Background paper commissioned by ECAPAPA under the ASARECA-IFPRI project Strategies and Priorities for Sub-Regional Agricultural Development and Agricultural Research-for-Development in Eastern and Central Africa. Entebbe: ECAPAPA.

Nkonya, E., J. Pender, P. Jagger, D. Sserunkuuma, C. Kaizzi, and H. Ssali. 2004. Strategies for sustainable land management and poverty reduction in Uganda. Research Report 133. Washington, D.C.: International Food Policy Research Institute.

Omamo, S. W. 1998a. Farm-to-market transaction costs and specialization in small-scale agriculture: Explorations with a nonseparable household model. *Journal of Development Studies* 35 (2): 152–163.

———. 1998b. Transport costs and smallholder cropping choices: An application to Siaya District, Kenya. *American Journal of Agricultural Economics* 80 (1): 116–123.

———. 2002. Efficiency and equity in public investment in agriculture: Lessons from soil fertility research in Kenya. In *Perspectives on agricultural transformation: A view from Africa*, ed. T. Jayne, I. Minde, and G. Argwings-Kodhek. Hauppauge, N.Y., U.S.A.: Nova Science Publishers.

Owango, M., B. Lukuyu, S. J. Staal, M. Kenyanjui, D. Njubi, and W. Thorpe. 1998. Dairy co-operatives and policy reform in Kenya: Effects of livestock service and milk market liberalization. *Food Policy* 23 (2): 173–185.

Pearson, S. R., C. P. Timmer, and W. P. Falcon. 1983. *Food policy analysis.* Baltimore: Johns Hopkins University Press.

Pederson, P. 2000. The changing structure of transport under trade liberalization and globalization and its impact on African development. CDR Working Paper 98.16. Copenhagen: Centre for Development Research.

Pender, J. 2004. Development pathways for hillsides and highlands: Some lessons from Central America and East Africa. *Food Policy* 29 (4): 339–467.

Pender, J., F. Place, and S. Ehui. 1999. Strategies for sustainable agricultural development in the east African highlands. EPTD Discussion Paper 41. Washington, D.C.: International Food Policy Research Institute.

———, eds. 2006. *Strategies for sustainable land management: Lessons from the east African highlands.* Washington, D.C.: International Food Policy Research Institute.

Pender, J., P. Jagger, E. Nkonya, and D. Sserunkuma. 2004. Development pathways and land management in Uganda. *World Development* 32 (5): 767–792.

Poulton, C., P. Gibbon, B. Hanyani-Mlambo, J. Kydd, W. Maro, M. Nylandsted Larsen, A. Osorio, D. Tschirley, and B. Zulu. 2004. Competition and coordination in liberalized African cotton market systems. *World Development* 32 (3): 519–536.

PRAPACE (Potato and Sweet Potato Improvement Program in Eastern and Central Africa). 2003. Setting priorities in ASARECA's networks, programs and projects—PRAPACE. Kampala: Regional Potato and Sweetpotato Improvement Network in Eastern and Central Africa.

RABESA (Regional Approach to Biotechnology and Biosafety Policy in Eastern and Southern Africa). 2004. The RABESA initiative—Towards a regional approach to biotechnology and biosafety policy in eastern and southern Africa. Entebbe: ECAPAPA.

RATES (Regional Agricultural Trade Export Support). 2005. Regional Agricultural Trade Export Support (RATES) Program. <www.ratescenter.org>.

RATIN (Regional Agricultural Trade Intelligence Network). 2005. Regional Agricultural Trade Intelligence Network. <www.ratin.net>.

Sabel, C. 1994. Learning by monitoring: The institutions of economic development. In *The handbook of economic sociology*, ed. N. Smelser and R. Swedberg. Princeton, N.J., U.S.A.: Princeton University Press.

SACRED-Africa (Sustainable Agriculture Centre for Research and Development in Africa). 2005. Sustainable Agriculture Centre for Research and Development in Africa. <http://www.acts.or.ke/sacred/>.

SADC (Southern African Development Community). 2004. Dar es Salaam declaration contains pro-active strategy for regional food security. 2004. *SADC Today* 7 (June). <http://www.sardc.net/editorial/sadctoday/v7-2-6-04/fsecurity.htm>.

SAGE (Centre for Sustainability and the Global Environment). 2004. Global land use database. University of Wisconsin, Madison. <http://www.sage.wisc.edu:16080/iamdata/>.

Sengooba, T. 2006. Status report: GM crop field trial and commercialization activities on the African continent. Slide presentation. Kampala: Program for Biosafety Systems Eastern Africa Office, Kampala.

Seward, P., and D. Okello. 2000. SCODP's mini-pack method: The first step towards improved food security for small farmers in western Kenya. Paper presented at the 68th Annual Conference of the International Fertilizer Industry Association, 22–25 May, Oslo.

Staal, S., and W. N. Kaguongo. 2003. The Ugandan dairy sub-sector: Targeting development opportunities. A strategic criteria for rural investment in productivity report. Washington, D.C.: International Food Policy Research Institute. <http://www.foodnet.cgiar.org/scrip/docs&databases/scrip_II_outputs2001/index.htm>.

Tomich, T., P. Kilby, and B. Johnston. 1995. *Transforming agrarian economies: Opportunities seized, opportunities missed.* Ithaca, N.Y., U.S.A.: Cornell University Press.

UNICEF (United Nations Children's Fund). 2001. *Global data base on child malnutrition 2001.* <http://www.childinfo.org/eddb/malnutrition/database1.htm>.

United Nations. 2005. *United Nations human development report 2005.* <http://hdr.undp.org/reports/global/2005>.

UNPD (United Nations Population Division). 1999. *World population prospects: The 1998 revision.* New York: United Nations.

Vogel, S. J. 1994. Structural changes in agriculture: Production linkages and agricultural demand-led industrialization. *Oxford Economic Papers,* new series 46 (1): 136–156.

Wood, S., K. Sebastian, F. Nachtergaele, D. Nielsen, and A. Dai. 1999. *Spatial aspects of the design and targeting of agricultural development strategies.* EPTD Discussion Paper 44. Washington, D.C.: International Food Policy Research Institute.

Wood, S., L. You, and W. Baitx. 2001a. *DREAM user manual,* version 3. Washington, D.C.: International Food Policy Research Institute.

———. 2001b. *Dynamic research evaluation for management (DREAM).* Washington, D.C.: International Food Policy Research Institute.

Made in the USA
Monee, IL
05 November 2026

67379799R00090